Health Technologies and Informatics

Health Technologies and Informatics: Research and Developments provides a comprehensive overview of mobile health applications, biodata management and analytics, medical imaging, personalized and public health systems, and biosignal processing.

With a focus on medical informatics, which has been identified as a necessity relating to health challenges relevant to the current pandemic, this book highlights engineering applications and methodologies involved in data evaluations. Detailed information is provided on diseases which could be monitored and necessary intervention for the treatment of these diseases or medical conditions.

Features:

- Provides recent advances on research and developments in the field of biomedical and health informatics
- Introduces topics such as mobile messaging, objectified information exchange, SmartCare, IoT-driven healthcare, cybersecurity issues, AI-enhanced healthcare, and so forth
- Covers novel engineering applications and methodologies involved in the pertinent data evaluations
- Includes dedicated chapters on machine learning in management and mitigation of COVID-19
- Explores the role of extended reality in health care including virtual, augmented, and mixed reality

This book is aimed at researchers and graduate students in biomedical and computer engineering.

Health Technologies and Informatics

Research and Developments

Edited by

Charles Oluwaseun Adetunji

Segun Fatumo

Kingsley Nnanna Ukwaja

CRC Press is an imprint of the
Taylor & Francis Group, an **informa** business

First edition published 2025
by CRC Press
2385 NW Executive Center Drive, Suite 320, Boca Raton FL 33431

and by CRC Press
4 Park Square, Milton Park, Abingdon, Oxon, OX14 4RN

CRC Press is an imprint of Taylor & Francis Group, LLC

Library of Congress Cataloging-in-Publication Data
Names: Adetunji, Charles Oluwaseun, editor. | Fatumo, Segun, editor. |
Ukwaja, Kingsley Nnanna, editor.
Title: Health technologies and informatics : research and developments /
edited by Charles Oluwaseun Adetunji, Segun Fatumo, and Kingsley Nnanna
Ukwaja.
Description: First edition. | Boca Raton, FL : CRC Press, 2025. | Includes
bibliographical references and index.
Identifiers: LCCN 2024034543 (print) | LCCN 2024034544 (ebook) | ISBN
9781032313801 (hardback) | ISBN 9781032313825 (paperback) | ISBN
9781003309468 (ebook)
Subjects: MESH: Medical Informatics Applications | Artificial Intelligence
| Telemedicine
Classification: LCC R855.3 (print) | LCC R855.3 (ebook) | NLM W 26.55.A7
| DDC 610.285--dc23/eng/20241112
LC record available at https://lccn.loc.gov/2024034543
LC ebook record available at https://lccn.loc.gov/2024034544

ISBN: 978-1-032-31380-1 (hbk)
ISBN: 978-1-032-31382-5 (pbk)
ISBN: 978-1-003-30946-8 (ebk)

DOI: 10.1201/9781003309468

Typeset in Times
by SPi Technologies India Pvt Ltd (Straive)

Contents

About the Editors

Professor Charles Oluwaseun Adetunji is a faculty member in the Microbiology Department within the Faculty of Science at Edo State University Uzairue, Edo State, Nigeria. He employs biological techniques and microbial bioprocesses to advance sustainable development goals and foster agrarian revolution through quality teaching, research, and community development. He previously held roles as the Acting Director of Intellectual Property and Technology Transfer, the Head of the Department of Microbiology, Sub Dean of the Faculty of Science, immediate Dean of the Faculty of Science and is presently serving as the Chairman of the Grant Committee as well as the Director of Research and Innovation at EDSU. Professor Adetunji commenced his early career at the Nigerian Stored Product Research Institute (NSPRI) in Ilorin, Nigeria, as a Research Officer II and progressed to the position of Principal Scientist. He then transitioned to Landmark University, Omu-Aran, where he served as a lecturer in various capacities, engaging in teaching and research. Additionally, he served as a visiting fellow at the Hebrew University of Jerusalem, Israel, specializing in Food Safety, and visited the Wageningen Centre for Development and Innovation in the Netherlands.

He is a Fellow of the Royal Society of Biology (UK), the Biotechnology Society of Nigeria, the Nigerian Young Academy, and the Nigerian Society of Microbiology. Additionally, he holds visiting professorships at the Center of Biotechnology, Precious Cornerstone University, Ibadan, and Trinity University, Lagos State. He is an affiliate member of the African Academy of Science. Presently, he serves as an external examiner for numerous academic institutions globally, particularly for Ph.D. and MSc students, including the University of Johannesburg, South Africa, Covenant University, Ota Lagos State, Ahmadu Bello University, Zaira, Landmark University, Kwara State among others. He is an external assessor to the Department of Biotechnology at Addis Ababa University to assist in establishing prima facie of candidates to be promoted to the rank of Full Professor, as well as at the University of Ilorin for the Department of Microbiology, and many other universities worldwide. He was recently appointed as a Visiting Professor at the Center of Biotechnology, Federal University of Technology, Abeokuta, Ogun State, and as an adjunct Professor at The University of the People, Pasadena, CA, USA. Moreover, He was recently appointment as a member of the Advisory board for College of Agriculture and Natural Sciences by the management of Joseph Ayo Babalola University, Nigeria.

He has received numerous scientific awards and grants from prestigious academic bodies such as the Council of Scientific and Industrial Research (CSIR) India, the Department of Biotechnology (DBT) India, The World Academy of Science (TWAS) Italy, Netherlands Fellowship Programme (NPF) Netherlands, The Agency for International Development Cooperation in Israel, Royal Academy of Engineering in the UK, and the National Research Fund (TETFUND), among others.

Due to his international recognition and significant contributions to knowledge and scientific impact, Professor Adetunji was recently listed in Stanford University's World Top 2% Scientists Rankings for 2023 while he was also recognized in the prestigious 2024 Career-long List of the World's Top 2% Scientists, as well as the single-year Most Impactful List of Distinguished Professors globally. He has authored numerous scientific journals and conference proceedings in refereed national and international journals amounting to over 812 manuscripts, including more than 60 books, 537 book chapters, and numerous scientific patents while he had also published many manuscripts in top rated journal such as Lancet with impact factor of 168.9. Notably, he holds a Google Scholar of i10-Index of 207. Additionally, he was recently ranked first among the top 500 prolific authors in Nigeria between 2020 till date by SciVal/SCOPUS in Elsevier, Netherland. Moreover, he achieved the top ranking in two distinct fields, Biological Sciences and Agricultural Science, among 500 prolific authors from 58 different countries in Africa in the year 2023 till date. Currently, he serves as a book series editor with Wiley and Sons, Elsevier, Springer, Taylor and Francis, among others.

Professor Adetunji has participated in over 100 international scientific conferences and workshops, assuming various roles such as Invited Speaker, Plenary Speaker, Keynote Speaker, Chief Panelist, Lead Guest Speaker, Chairman of Scientific Sessions, and Judge of Posters, across several continents worldwide. Additionally, he serves as a reviewer for more than 100 scientific journals, including Nature, Communications Earth & Environment in Nature, PLOS One, among others. His contributions to open science and the role his research plays in addressing global challenges were recently acknowledged by Elsevier, particularly after some of his scientific articles were associated with the United Nations (UN) Sustainable Development Goals (SDGs).

Currently, Professor Adetunji holds the positions of President and Chairman of the Governing Council of the Nigerian Bioinformatics and Genomics Network Society. Previously, he had served as the General Secretary of the Nigerian Young Academy, previously chairing the Biological and Health Sciences Working Group for the Nigerian Young Academy. Furthermore, he holds the positions of Assistant General Secretary and National Coordinator for the South-West Nigerian Society for Microbiology. He was recently appointed to the Young Investigator Award Committee by the American Society of Tropical Medicine and Hygiene, USA. He also serves as the Vice-President and Nigerian Country Director of the Genetics Society of West Africa and as the Director of the International Affiliation and Training Centre for Environmental and Public Health, Research, and Development, situated in Zaria. Moreover, he acts as the immediate past Business Manager and currently the Vice-President for the Biotechnology Society of Nigeria. Prof Adetunji was appointed by the African Union Development Agency-NEPAD Centre of Excellence in Science, Technology, and Innovation (CoE STI) as a senior expert and consultant to facilitate the development of curricula in Biotechnology, Applied Microbiology, Microbiome, and Genome Editing in various African countries, including Nigeria, Ethiopia, Ghana, Zimbabwe, Mozambique, Malawi, Kenya, among others. He was recently appointed as the Vice Chairman of Technical Expert Committee by Standard Organization of Nigeria on Sugar, Confectionery, Biscuits, and Bakery Products.

In recognition of his expertise and extensive experience in the field of Microbiology, Bioinformatics and Biotechnology, Professor Adetunji holds the unique privilege of serving as a Consultant to the Government of the Republic of Serbia in Europe, specifically as a Project Peer Reviewer for the Science Fund in the areas of Microbiology, Nanotechnology, Food Science, and Biotechnology. Also, he was recently appointed as a senior expert by South Africa's National Research Foundation (NRF) in evaluating the quality, impact and standing of the research portfolio of Scientists in the area of Microbiology that are associated with a Higher Education/ Research Institution (HEI) in South Africa (SA). Moreover, he was recently appointed to serve as a reviewer of DSI/NRF South African Research Chair Initiative Programme (SARChI) - Performance of Grant Recipients received by distinguished professors in the area of Biotechnology Innovation & Engagement in South African. Recently, he was appointed as a member of College of Experts from UK to serve as a member of selection committee on Africa Research Excellence Fund to help in timely delivery on its vision of an inspired, committed, and talented community of researchers in Africa leading world-class research and participating equitably in international research endeavors for health and wellbeing. Additionally, he was recently appointed as the general coordinator and supervisor for the 1000 scientific trials programme by Harvest Harmonics in the USA. In this role, he oversees coordination across 58 countries in Africa, serving as the Country Representative for Nigeria. This appointment underscores the significance of his landmark publications in translational research, his diligent work ethic, and exceptional talents, positioning him as a renowned trailblazer in advancing academics and research in Africa.

On a global scale, Professor Adetunji has mentored numerous undergraduate, postgraduate, and post-doctoral students at both national and international universities, including Jawaharlal Nehru University, India, among others. He has provided advice to numerous scientists and serve as a mentor to young researchers on how they could design translational and innovative research. Based on his experience as director of intellectual property at EDSU he has assisted many scientists on how they could file their innovation as well as assisted numerous foundations, specifically, on how to secure funding, grant, scholarship, and awards from non-profit organizations to assist qualified young Nigerians who aspire to pursue science-based disciplines at the advanced levels (Master/Ph.D) and assist them to get tuition fees and research fund.

Professor Adetunji is esteemed as an accomplished academic, erudite scholar, and researcher par excellence, boasting an impressive academic record and enviable wealth of experience in microbiology, biotechnology, and bioinformatics. He is the convener of the Recent Advances in Biotechnology conference, held annually since 2020, where distinguished Microbiologists and Biotechnologists convene to share their latest discoveries.

Prof. Segun Fatumo is the head of NCD Genomics at the MRC/UVRI & LSHTM Uganda Research Unit and Chair of Genomic Diversity at Queen Mary University of London and. He specialises in NCD Genomics and Omics of African Populations. He co-led the first GWAS of cardiometabolic traits in Africa and led the first GWAS of Kidney functions in continental African populations. He is the director of the KidneyGenAfrica Research Partnership Programme - A Partnership to Deliver Research and Training Excellence in Genomics of Kidney Disease in Africa.

Segun Fatumo is strongly committed to increasing diversity in genomic studies and was recently awarded the prestigious MRC Impact prize for advocating for the inclusion of Africa in genomic research and championing genetic risk prediction of complex diseases in Africa populations. He is a Fellow of the Higher Education Academy (FHEA) and is actively involved in capacity building across Africa.

Segun Fatumo received postdoctoral training in genetic epidemiology at the University of Cambridge and Wellcome Sanger Institute and a postdoctoral fellowship in Bioinformatics at the University of Georgia, Athens, USA. Prior to that, He had postgraduate training in Bioinformatics at the University of Cologne, Germany and Ph.D. in Computer Science (Bioinformatics specialization) from Covenant University, Nigeria.

Dr Kingsley Nnanna Ukwaja is a public health physician, epidemiologist, and sub-specialist trainee in clinical respiratory medicine with excellent experience in operational research and communicable and non-communicable disease epidemiology. Other interdisciplinary interests lie at the nexus of translational epidemiology, health economics, and implementation science. He is a highly motivated professional with over a decade of clinical and public health experiences in the Nigerian health care system. He has over ten years of programmatic management related to tuberculosis (TB) and HIV/AIDS at national and sub-national level. He has experience in the implementation of international (WHO) and national (in Nigeria) TB prevention, care, and treatment programs, programmatic management of TB and drug-resistant TB, research synthesis, economic evaluation, policy development, monitoring and evaluation, supportive supervision, and operational research. In the field of translational epidemiology he has a strong interest in creating dynamic modeling analyses and frameworks that translates epidemiological data into effective decisions. Early work on TB program data centered on translating epidemiological data to useful practical and policy decisions. He has received both training and mentorship from the Pan-African Thoracic Society's Methods in Epidemiologic Clinical and Operations Research (PATS-MECOR) courses, and he now serves as an associate faculty and a role model to PATS-MECOR students from Africa. His other experiences include proposal writing, data management and analysis (Epi Info, SPSS, STATA), and manuscript/report writing.

He specializes in general medicine, public health, respiratory diseases, infectious diseases, health economics, and epidemiology.

Contributors

Adedeji Olufemi Adetunji
Laboratory of Immuno-nutrition
North Carolina A&T State University
Greensboro, North Carolina

Charles Oluwaseun Adetunji
Applied Microbiology, Biotechnology and Nanotechnology Laboratory
Department of Microbiology, Edo State University
Uzairue, Iyamho, Nigeria

Juliana Bunmi Adetunji
Department of Biochemistry, Faculty of Basic Medical Science
Osun State University
Osogbo, Nigeria

Adebayo Olusola Adetunmbi
Department of Computer Science
Federal University of Technology
Akure, Ondo State, Nigeria

Olorunsola Adeyomoye
Department of Physiology
University of Medical Sciences
Ondo City, Nigeria

Akinmoju Olumide Damilola
Department of Paediatrics
Ladoke Akintola University of Technology Teaching Hospital
Ogbomosho, Oyo State, Nigeria

K. I. T. Eniola
Department of Biological Sciences
Joseph Ayo Babalola University
Ikeji Arakeji, Nigeria

Hannah Edim Etta
Animal and Environmental Biology, Faculty of Biological Sciences
Cross River University of Technology
Calabar, Nigeria

Aishatu Idris Habib
Applied Microbiology, Biotechnology and Nanotechnology Laboratory
Department of Microbiology, Edo State University
Uzairue, Nigeria

Onyijen Ojei Harrison
Department of Mathematical and Physical Sciences
Samuel Adegboyega University
Ogwa, Nigeria

Olusanmi Ebenezer Odeyemi
Department of Science Laboratory Technology
Federal College of Animal Health and Production Technology
Ibadan, Nigeria

Oluwakemi Mary Odeyemi
Department of Physics, College of Natural Sciences
Joseph Ayo Babalola University
Ikeji Arakeji, Nigeria

Frank Abimbola Ogundolie
Department of Biotechnology
Baze University
Abuja, Nigeria

Onotu Roseline Ohunene
Department of Paediatrics
Federal Medical Centre
Jabi, Nigeria

Michael O. Okpara
Department of Biochemistry
Federal University of Technology
Akure, Nigeria

Abdulrahmon A. Olagunju
Department of Human Physiology
The Federal University of Technology
Akure, Nigeria

Olugbemi T. Olaniyan
Laboratory for Reproductive Biology and Developmental Programming
Department of Physiology, Rhema University
Aba, Nigeria

Akinola Samson Olayinka
Computational Science Research Group, Department of Physics
Edo State University
Uzairue, Nigeria

Tosin Comfort Olayinka
Department of Information Technology & Cyber Security
College of Computing, Wellspring University
Benin City, Nigeria

Ale Oluwabusolami
Department of Paediatrics
Federal Medical Centre
Jabi, Nigeria

Clement Atachegbe Onate
Department of Physical Sciences, College of Pure and Applied Sciences
Landmark University
Omu-Aran, Nigeria

Oluwafemi Adebayo Oyewole
Department of Microbiology
Federal University of Technology
Minna, Nigeria

Tolulope Peter Saliu
Department of Biochemistry
Federal University of Technology Akure
Nigeria.

Okanlawon Taiwo Stephen
Department of Biological Sciences
Samuel Adegboyega University Ogwa
Edo State, Nigeria

Olotu Titilayo
Department of Microbiology
Adeleke University
Ede Osun State, Nigeria

Igiku Victory
Applied Microbiology, Biotechnology and Nanotechnology Laboratory
Department of Microbiology, Edo State University
Uzairue, Iyamho, Nigeria

Nyejirime Young Wike
Department of Human Physiology, Faculty of Basic Medical Science
Rhema University
Aba, Nigeria

Preface

The whole world is currently going through difficult times as a result of the COVID-19 pandemic and uncountable health challenges which have claimed the lives of many people. The situation is worse in remote areas in developing countries where there is no immediate access to healthcare facilities. This has led to several health emergencies as a result of several diseases outbreaks such as COVID-19, Lassa fever, Ebola, and many more. Most of time it takes several hours before the healthcare personnel can reach the affected communities due to poor road networks as well as a lack of internet facilities. Through prediction modeling utilizing artificial intelligence, machine learning algorithms, and mathematical approaches many epidemic and pandemics diseases have been forecasted and predicted such as Zika virus, HIV/AIDS, Ebola virus, COVID-19, SARS, and MERS. The ability to rapidly predict potential epidemics will produce better control measures towards the outbreaks for improved early detection, early analysis, interpretation, amplification, and eradication. Therefore, there is a need for mobile healthcare stations for coordination, circulation, and transfer of necessary information between the affected communities with necessary healthcare offices for their immediate response. Furthermore, the application of medical informatics plays a crucial role in the utilization of computer technology by numerous health professionals in various fields such as genomics, microbiology, physiology, telemedicine, medical robotics, application of ICT, web-based medical education or distance learning The role of machine learning in addressing the challenges associated with detection, diagnosis, and management of COVID-19 as well as classification, drug discovery, and prognosis is presented in this book. The mode of transmission of COVID-19 disease and variants of the disease are presented. Recent advances in machine learning algorithms for prognosis, prediction, diagnosis, treatments, and managements of the COVID-19 disease are discussed. This book also provides detailed information on the application of artificial intelligence in dentistry and orthodontics as well as to provide a detailed information on the application of artificial intelligence and the deep learning process and drug discovery. This book also highlights the advantages of SmartCare in biomedical technologies and health informatics. This book also provides detailed information on recent advancement in genomics and understanding of the genetic basis of diseases have made personalized medicine an emerging medical practice for easier diagnosis and treatment. This book also provides detailed information on the recent advances in the application of IoT, wearable technologies, and robotics to healthcare delivery. Insights are given on existing areas of IoT, wearable and robotics applications as well as recent success stories on its implication on healthcare delivery. Specific cases of applications to diverse diseases which affect human life are presented with possible research areas for future developments in the field. This book also provides detailed information on the role of 5G wireless technology in the provision of sustainable healthcare services at minimal cost and improved efficiency. A comprehensive examination of the evolution of wireless networks from 1G to 6G is presented with their strengths and weakness. The various applications of the 5G network in healthcare delivery are presented as well as some of the challenges that might impair its usage. This book is aimed at a diverse audience, including the health sectors, federal and state ministries of health, policymakers, researchers and governments and scientists in different fields such as microbiology, medical health care, virology, bioinformatics, epidemiology, computer science, chemical pathology, computational biology, and biomedical engineering, and many more are highlighted. Furthermore, I want to thank my coeditors for their effort and dedication during this project. Moreover, I wish to gratefully acknowledge the suggestions, help, and support of Dr. Gagandeep Singh who is the Senior Publisher (STEM) CRC Press India and Sub-Saharan Africa under Taylor & Francis Books India Pvt Ltd.

Professor Charles Oluwaseun Adetunji
Ph.D, AAS affiliate, FRSB (UK) FNYA; FBSN; FNSM, MNBGN
Director of Research and Innovation, Edo State University, Uzairue, Nigeria
December 2023

1 Relevant Application of Machine Learning in the Management of COVID-19

Tosin Comfort Olayinka
Wellspring University, Benin City, Nigeria

Adebayo Olusola Adetunmbi
Federal University of Technology, Akure, Nigeria

Akinola Samson Olayinka and Charles Oluwaseun Adetunji
Edo State University, Uzairue, Nigeria

Onyijen Ojei Harrison
Samuel Adegboyega University, Ogwa, Nigeria

Okanlawon Taiwo Stephen
Samuel Adegboyega University Ogwa, Edo State, Nigeria

INTRODUCTION

Despite the rapid development of the COVID-19 vaccine and hastened administration of the different varieties of the vaccine across the world, the pandemic is still a major health concern globally. The present statistics from the World Health Organization (WHO) show that over 1.5 million new cases are being reported daily while the total confirmed cases is around 350 million with the confirmed deaths of over 5.5 million globally (WHO 2022). Research frontiers from various disciplines have made significant contributions to how to reduce the impact of this deadly disease globally. Machine learning (ML), which is an offshoot of artificial intelligence (AI), techniques have been applied I, several areas such as road accident prediction, control and management (Ghandour et al. 2020; Gianfranco et al. 2017; Gutierrez-Osorio & Pedraza 2020; Nwankwo et al. 2022; Rajkumar et al.

DOI: 10.1201/9781003309468-1

2020); agricultural related areas of food production and preservation (Anani et al. 2022; Olayinka et al. 2022); material synthesis, discovery and properties prediction (Iwasaki et al. 2019; Liu et al. 2021; Olayinka et al. 2020; Pollice et al. 2021; Raccuglia et al. 2016; Tao et al. 2021). In the healthcare system, machine learning has played a key role in the fight against the COVID-19 outbreak. Several studies have reported COVID-19 prediction, prognosis, identification, diagnosis, risk profiling, and management using machine learning techniques (Bertsimas et al. 2020; Punn et al. 2020; Tschoellitsch et al. 2021; Wu, Zhang, et al. 2020). Healthcare systems globally are embracing AI as well machine learning techniques to improve healthcare delivery. With progress in areas like Internet of Medical Things (IoMT), telemedicine, and other application of AI in medicine, the future of healthcare delivery and management is closely connected to machine learning and artificial intelligence. This chapter presents a basic overview of SARS-CoV-2, generally known as COVID-19, its origin and variants up to date as well as its management using machine learning techniques.

The Origin of the COVID-19 Pandemic

As far back as 2007, Cheng et al. (Cheng et al. 2007) had raised a serious concern about the coronavirus being "an agent of emerging and reemerging infection." Severe acute respiratory syndrome (SARS) coronavirus (SARS-CoV) was first recorded in 2003 (Cheng et al. 2007). The SARS-CoV-2 is a beta(β)-coronavirus, which is an enveloped non-segmented positive-sense RNA virus. SARS-CoV-2 was first discovered in patients with pneumonia in Wuhan, China, in 2019, and it's generally referred to as COVID-19 (Morens et al. 2020; Shereen et al. 2020; Zhou et al. 2020; Zhu et al. 2020; Hu et al. 2021; Platto et al. 2021). The virus is divided into five genera, including alpha (α-), beta (β-), gamma (γ-), delta (δ-), and omicron (o-) coronavirus. The α- and β-coronavirus are able to infect mammals, while γ- and δ-coronavirus tend to infect birds. Six strains of coronavirus were suspected as human-inclined virus, among which α-coronavirus and β-coronavirus with low pathogenicity cause moderate respiratory signs similar to a common cold. The specific recognized β-coronavirus, SARS-coronavirus and MERS-coronavirus lead to severe and possibly fatal respiratory diseases (Yin & Wunderink 2018).

COVID-19 has been described by the Centre for Disease Control (CDC) as a disease that is caused by one strain of SARS-CoV-2 recognized globally, called the unconventional coronavirus. COVID-19 was identified through an outbreak in Wuhan, China. The symptoms associated with it can have effect on the human lungs and respiratory system with well-known symptoms such as cough, an immoderate temperature, and shortness of breath.

It was determined that the genome series of SARS-CoV-2 is about 96% similar to a bat-related CoV RaTG13, meanwhile it accounts for about 79% identification of SARS-coronavirus. With references to the analysis conducted on genomics, the bat has been recognized as the origin of the virus, and SARS-CoV-2 is transmitted from bats through unidentified intermediate hosts to contaminate human beings (Zhou et al. 2020). The five variants of concerns identified so far and where they were first discovered are shown in Figure 1.1. Table 1.1 shows types, source, and the nature of

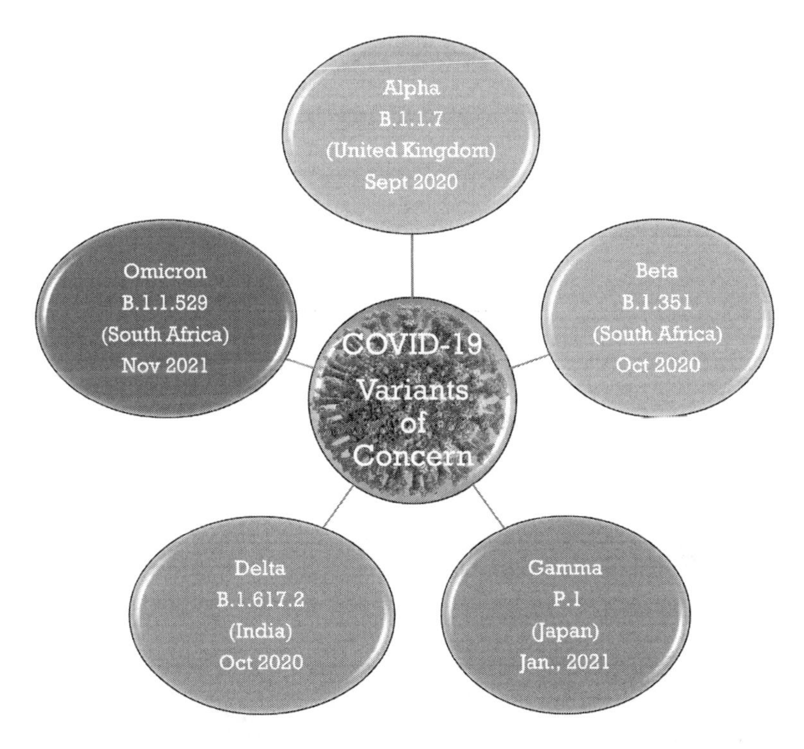

FIGURE 1.1 COVID-19 variants of concern.

TABLE 1.1
Coronavirus Types, Natural Reservoir and the Nature of Illness Caused

Type	Coronavirus	Host	Nature of Illness	Reference
Alpha	HCoV-NL63 HCoV-229E	Mammals e.g., Bats	Mild	CDC (2022a)
Beta	SARS-CoV SARS-Cov-2 MERS-CoV HCoV-OC43	Mammals e.g., Bats	Severe and mild	
Gamma	Gamma-CoV	Birds	Severe	
Delta	Delta	Birds	More severe and more contagious	CDC (2022b)
Omicron	B.1.1.7 SARS-CoV-2	—	Highly transmissible with several mutations	UNICEF (2022)

illness caused by different variants of coronavirus. Types of coronavirus variants and associated illness as well as its origin and date of discovery are presented in Table 1.2. Figure 1.2 shows both the viral and host factors that influence the transmission of SARS-CoV-2. Viral factors are the S protein, M protein, N, E protein and cofactors

TABLE 1.2
Coronavirus Variants, Associated Illness and Its Origin

Human Coronavirus Name	Diseases or Illnesses Caused	Origin	Year
SARS-Cov-2	COVID-19	Wuhan, China	2019
MERS-CoV	Middle East respiratory syndrome	Saudi Arabia	2012
SARS-CoV	Severe acute respiratory syndrome	Guangdong, China	2002
HCoV-NL63	Mild respiratory diseases	Netherlands	2004
HCoV-229E		USA	1960s
HCoV-OC43		China	1967
HKU1		Hong Kong	2004

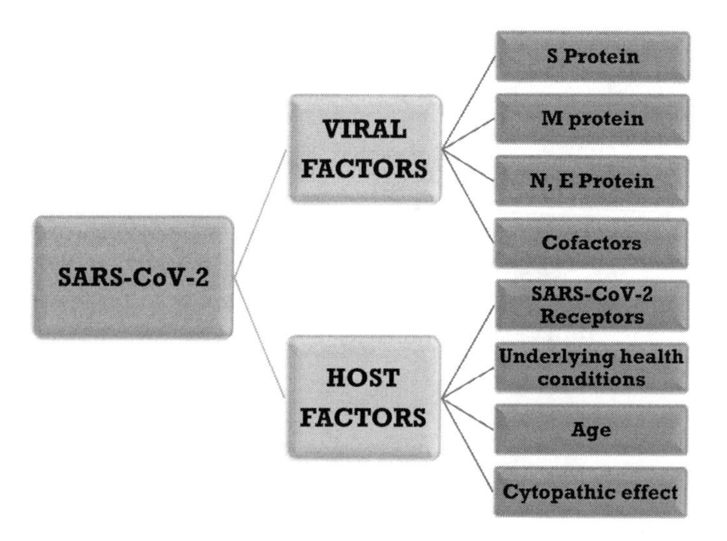

FIGURE 1.2 Host and viral factors that influence the transmission of SARS-CoV-2.

while the host factors include SARS-CoV-2 receptors, fundamental health situations, age, and cytopathic effect (CPE).

Mode of Transmission

The SARS-coronavirus and MERS-coronavirus is said to arise particularly through hospital transmission. Healthcare workers were infected in 33–42% of SARS cases, and transmission between patients (62–72%) was the most common source of infection in MERS-CoV cases (Chowell et al. 2015). SARS-CoV-2 is primarily transmitted through direct contact with transitional host animals or consumption of infected wild animals.

However, the supply(s) and transmission habitual(s) of SARS-CoV-2 remain elusive. SARS-CoV-2 secondary transmission from human to human happens usually within family members or close relatives and friends who in detail contacted with patients or incubation vendors. It is said that 31.3% of sufferers recently travelled to Wuhan and 72.3% of sufferers had contact with people from Wuhan; many of the sufferers were non-citizens of Wuhan (Guan et al. 2020).

The main source of extensive sorts of coronaviruses are bats, along with intense acute respiration syndrome coronavirus (SARS-CoV)-like viruses. Virus-host interactions affect viral access and replication. SARS-CoV-2 is an enveloped high-quality unmarried-stranded RNA (ssRNA) coronavirus. Two-thirds of viral RNA, especially placed within the first open reading body (ORF 1a/b), encodes 16 non-shape proteins (NSPs).

The other segment of the genomic content of the virus encodes four vital structural proteins, namely; nucleocapsid (N) protein, small envelope (E) protein, spike (S) glycoprotein, and matrix (M) protein, as well as several accessory proteins. The crucial step for virus entry occurs when the S glycoprotein of SARS-CoV-2 binds to host cell receptors, angiotensin-changing enzyme 2 (ACE2). The feasible molecules facilitating membrane invagination for SARS-CoV-2 endocytosis are nonetheless uncertain. There are some proteins that increase pathogenesis of the virus. Host factors (lower panel, Figure 1.2) can also influence susceptibility to infection and disease progression. The elderly and individuals with underlying health conditions are at risk of SARS-CoV-2 infection and are more likely to develop severe illness.

Epidemiology – Reservoirs and Transmission

The epidemic of unknown acute breathing tract contamination started in Wuhan, China, on 12 December 2019, and it was probably connected to a market where seafood is sold. Several researchers have pointed out that bats could be the source of SARS-CoV-2 (Calisher et al. 2006; Giovanetti et al. 2020, 2021; Irving et al. 2021). However, there is no concrete evidence so far that the origin of SARS-CoV-2 was the seafood marketplace. Bats are the natural source for a wide variety of coronaviruses, including SARS-CoV-like and MERS-CoV-like viruses. Upon virus genome sequencing, COVID-19 was analyzed at some point of the genome to Bat CoV RaTG13 and confirmed 96.2% standard genome series identification (Wu, Zhao et al. 2020).

Besides, protein sequences alignment and phylogenetic investigation revealed a related residue of receptor have been discovered in lots of species, which suggest other possible transitional hosts, which include pangolin, turtles, and snacks. SARS-CoV-2 human-to-human transmission takes place especially between circles of relatives, inclusive of spouse and children and associates who have close contact with patients or incubation providers.

Transmission among healthcare workers occurred in 3.8% of COVID-19 patients, as reported by the National Health Commission of China on 14 February 2020. By evaluation, the transmission of SARS-CoV and MERS-CoV is said to arise specifically via nosocomial transmission. Infections of healthcare workers in 33%–42% of SARS instances and transmission between patients (62–79%) became the maximum common route of contamination in MERS-CoV instances (Chowell et al. 2015). Contact with transitional host animals or intake of wild animals became the suspected main course of transmission of SARS-CoV-2; however, the supply(s) and transmission ordinary(s) of SARS-CoV-2 continue to be elusive (Kang et al. 2017).

Omicron SARS-CoV-2 Variant

On November 25, 2021, about 23 months after the first stated case of COVID-19 and after an estimated 260 million cases internationally and 5.2 million deaths (WHO, 2021), a new wave SARS-CoV-2 variant of concern (VoC), Omicron become evident. Omicron appeared in a COVID-19-ravaged world where people were worn-out and frustrated due to the devasting effect of the pandemic on social, intellectual, moral, and economic life. Although the preceding coronavirus emerged in a world wherein vaccination from COVID-19 infections become commonplace, this fifth variant of coronavirus appeared at a time while vaccine immunity is increasing inside the world.

The appearance of alpha, beta, and delta variants of SARS-CoV-2 VoCs had been related to novel waves of infections, occasionally throughout the world. For instance, the expanded transmissibility of the delta strain is related to a higher viral load, longer length of infectiousness, and high risk of reinfection, due to its potential to get away from natural immunity which resulted in the delta strain rapidly becoming the globally dominant version (Fontanet et al. 2021) The delta strain maintains new waves of contamination and stays the dominant strain for the duration of the fourth wave in many nations. Concerns about reduced vaccine efficiency due to new versions have modified our knowledge of the COVID-19 endgame, disabusing the sector of the notion that worldwide vaccination is in itself adequate for controlling SARS-CoV-2 infection (Townsend et al. 2021).

Economic Loss Related to COVID-19 and the Impact on Social Lifestyles

The COVID-19 pandemic affected the worldwide financial system in two ways. One, the spread of the virus endorsed social distancing, which caused the shutdown of monetary markets, company offices, corporations and activities. Two, the speed at which the virus spread, and the heightened uncertainty about how horrific the state of affairs could get, led to flight to safety in consumption and investment among clients and investors (Ozili & Arun 2020). In April 2020, the United States experienced a rapid surge in COVID-19 cases and deaths. Globally, the virus has infected more than 9.4 million people, with about one-third of these cases occurring in the United States, resulting in over 400,000 deaths. Over 80 countries closed their borders to contain the spread of the virus. Businesses, industries, and companies were ordered to close, and the public was advised to stay indoors, with strict enforcement of government regulations and penalties in countries such as Spain, France, India, and others.

The outbreak of this pandemic has impacted the social, political, economic, and religious systems of the world. The world's leading commercial economies such as the United States, Italy, France, and so forth are on the verge of exhaustion because of the overpowering effect and unpredictable effect of the pandemic. The restrictions imposed to manipulate the pandemic have had a devastating financial impact across the globe, leading to massive process losses and reduced incomes. Indeed, the raging debate (now not only in Africa) is on the way to stability the fitness blessings of imposed regulations and the economic costs of those regulations. The UN Economic Commission for Africa forecast that the combined impact of the crisis would result in a growth rate of 1.1% in 2020 under the best-case scenario and a contraction of

–2.6% under the worst-case scenario, depriving 19 million people of their livelihoods. In the context of weak social safety programs in Africa, this could push up to 29 million more people into poverty. Women and young people have been disproportionately affected by job losses and income declines across the region, as a significant portion of these businesses operate in the precarious informal and gig economies.

A statistical overview of COVID-19 reported cases in the top 15 countries globally and in the African continent is presented in Figures 1.3 and 1.4 respectively. The map distribution for the dataset presented in the chart is shown in the accompany map. Figure 1.3 shows the United States, India and Brazil as the top three with the highest cases of VOVID-19 while South Africa, Morocco, and Tunisia are the top three in the Africa continent as shown in Figure 1.4.

The top ten countries with the highest reported number of COVID-19 deaths globally and in the Africa continent are presented in Figures 1.5 and 1.6, respectively. Globally,

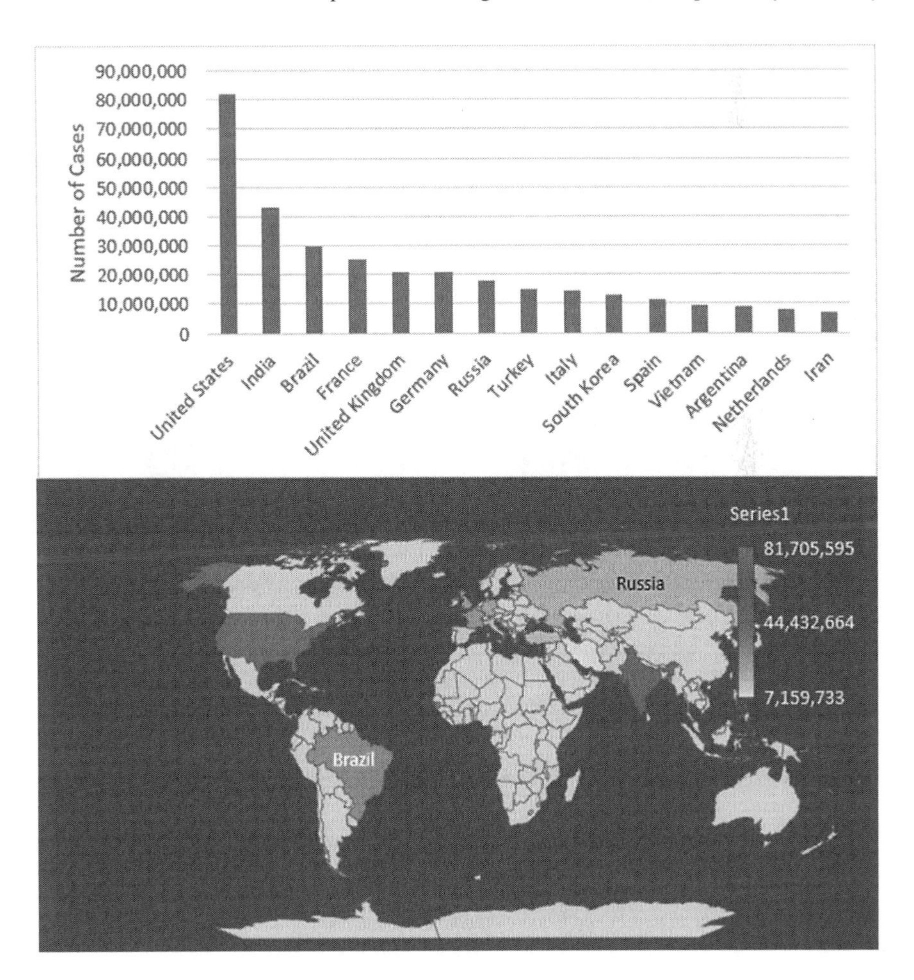

FIGURE 1.3 COVID-19 cases in the top 15 countries globally with highest reported cases as of March 2022.

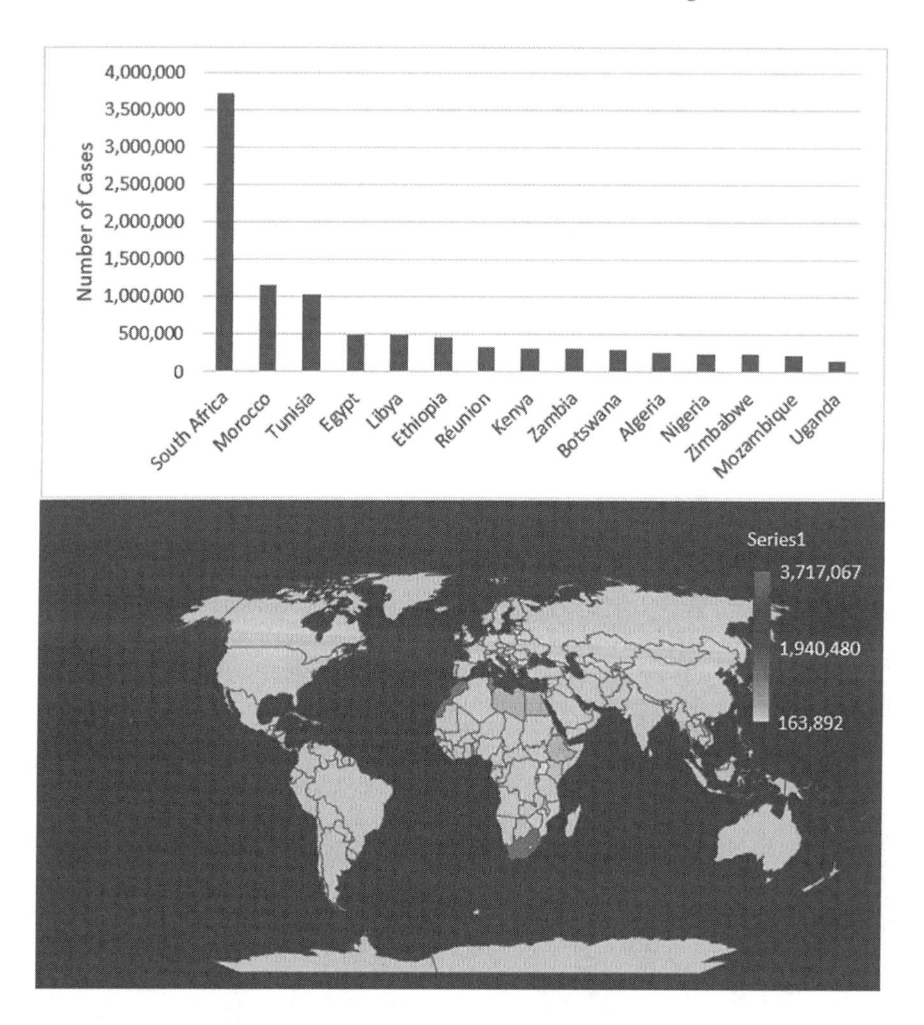

FIGURE 1.4 COVID-19 cases in the top 15 countries in Africa with the highest reported cases as of March 2022.

the top three countries with the highest number of deaths are the United States, India, and Brazil as shown in Figure 1.5 while South Africa, Tunisia, and Egypt are the top three with the highest number of COVID-19 deaths in Africa as shown in Figure 1.6.

Algorithms Used in Testing, Diagnosing, Managing, and Predicting COVID-19 Disease

Machine learning techniques are fast gaining ground for testing, diagnosing, managing, predicting COVID-19 disease, and treatments to enhance the results of patients. This study will highlight some of the work done using a machine learning approach to diagnose and predict COVID-19.

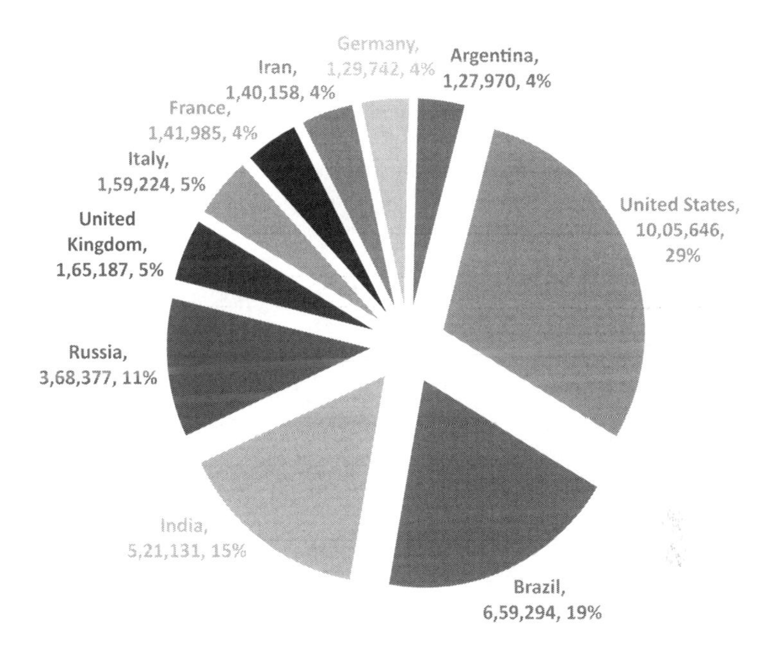

FIGURE 1.5 Ten countries with the highest reported COVID-19 deaths globally.

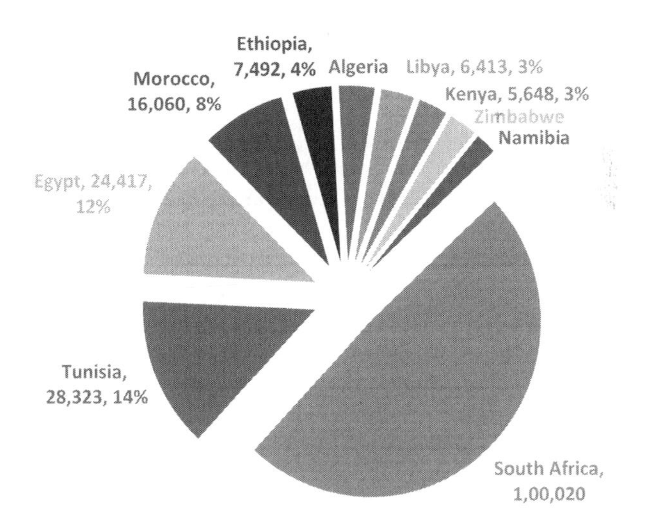

FIGURE 1.6 Ten countries with the highest reported COVID-19 deaths in Africa.

XGBoost Model

For Diagnosing

A classification model based on XGBoost was built to ascertain the disparity between influenza and COVID-19 patients (Li et al. 2020). The classification model was adapted to use data from symptoms and routine testing of patients to forecast the incidence of COVID-19. The COVID-19 dataset was acquired from published works from 151 articles and the data were subjected to analysis using a sample of 413 patients' clinical data. Their results showed that the patients' age, computed tomography scan (CT scan) results, body temperature of the subjects, lymphocyte levels, presence or absence of fever, and presence or absence of cough were the most significant features for their model development and prediction of COVID-19. The sensitivity and specificity of the prediction results were 92.5% and 97.9%, respectively.

Kukar et al. adopted Random Forest, XGBoost and deep neural networks (DNN) methods to predict COVID-19. The algorithms were utilized to create models that used attributes such as standard blood test results, age, and sex to predict COVID-19 diagnosis. The studies used data from 5333 patients that were admitted to the Infectious Diseases department of University Medical Centre, Ljubljana, out of which 160 were positive. The greatest results were achieved by XGBoost, which had an Area under the ROC (receiver operating characteristic) Curve (AUC) of 97%, a sensitivity of 81.9%, and a specificity of 97.9%. The model indicated that mean corpuscular hemoglobin concentration (MCHC), international normalized ratio (INR), and prothrombin activity percentage as well as eosinophil count and albumin were vital attributes in COVID-19 diagnosis and prediction (Kukar et al. 2021).

For Predicting

Vaid et al. (2020) created a classification model to predict deaths and critical state of patients as a result of COVID-19. The data collection was from five hospitals within New York City. A total of 4098 patients were sampled. The prediction model was built using XGBoost with baseline comparator models. For death prediction, AUCROC percentage of 0.89 at 3 days, 0.85 at 5 days and 7 days, while at 10 days it dropped to 0.84 for the model under consideration; AUCROC percentage reported for 3 days, 5 days, 7 days, and 10 days were 0.80, 0.79, 0.80 and 0.81 respectively from their model for critical event prediction. The study discovered that older age, anion gap, and CRP were critical factors for predicting death, while acute renal injury on admission, increased LDH, tachypnea, and hyperglycemia were the greatest effectors for predicting a critical event at 7 days.

Yan et al. created a model to predict COVID-19 patient survival. Tongji Hospital in Wuhan, China, provided a dataset of 404 infected individuals (213 survivors and 191 non-survivors). The XGBoost classifier was used in the study, which discovered that LDH, lymphocytes, and hs-CRP were the most discriminative indicators of patient survival. A model accuracy of greater than 90% was achieved (Yan et al. 2020).

Bertsimas et al. built the COVID-19 Mortality Risk tool, which predicts mortality using the XGBoost algorithm. The program was created utilizing data from 3927 COVID-19 positive patients from 33 institutions in Europe and the United States. Three validation cohorts were used to validate the model, with AUC values ranging from 0.92 to 0.81. The study revealed that increased age, decreased oxygen level, elevated CRP levels, blood urea nitrogen (BUN), and blood creatinine were the primary risk markers for mortality (Bertsimas et al. 2020).

Random Forest (RF) Model

Diagnosing

Wu et al. used RF to create an assisted discrimination tool that rapidly and accurately identified COVID-19 patients utilizing key blood indices from clinical blood test data. Multiple sources in China were used to compile the data collection, which included 253 samples from 169 suspected COVID-19 individuals. In addition, 27 individuals with confirmed COVID-19 instances provided 105 consecutive positive tests. To analyze the value of each component, the model development was done using 49 features at first. Eleven final input indications were chosen based on the top-ranking features. With an AUC of 99.26%, a sensitivity of 100%, and a specificity of 94.44% on an independent test set, the method's performance was comparable to that on the training set (Wu, Zhang, et al. 2020).

Tschoellitsch et al. (2021) used the Random Forest algorithm to construct a model that could predict the diagnosis of COVID-19 from regular blood testing. The model was developed using 1528 patients (65 positives), and it had an accuracy of 81%, an AUCROC of 0.74, a sensitivity of 60%, and a specificity of 82%. Leukocyte count, red blood cell distribution width (RDW), hemoglobin, and serum calcium were shown to be the most critical indicators in predicting diagnosis in the study.

Logistic Regression (LR) Model

For Diagnosing

Based on full blood count components and patient sex, Joshi et al. built an LR model to predict COVID-19 PCR positivity. The training of the model was carried out using 33 positive and 357 negative datasets from Stanford Health Care. The researchers used three CBC components (absolute neutrophil count, absolute lymphocyte count, and hematocrit) as well as male sex as model features. A total of 236 positive examples and 2052 negative cases were used in the validation process. The model had a C-statistic of 78%, a sensitivity range of 86% to 93%, and a specificity range of 35% to 55%. The algorithm may be limited to predicting positive cases, according to the authors, allowing for a 33% increase in effectively spent resources (Joshi et al. 2020).

Shoer et al. used nine simple survey questions to create a prediction model. The study employed data from a nationwide symptom survey that was completed by nearly two million people in Israel. A total of 43,752 adults were included in the study, with 498 of them self-reporting COVID-19 positivity. Age, gender, prior medical conditions, smoking habits, and self-reported symptoms such as fever, sore throat, cough, loss of taste or smell, and shortness of breath were among the survey questions. An LR approach was used to train the model, which resulted in an AUC of 0.737 (Shoer et al. 2021).

Tordjman et al. studied 400 patients (258 positive) from three French hospitals: Cochin Hospital in Paris, Ambroise Par'e Hospital in Boulogne, and Raymond Poincar'e Hospital in Garches. A binary LR algorithm was used in the study to create a scoring model that predicted the likelihood of a positive COVID-19 diagnosis. The AUC of the model was 88.9%, the sensitivity was 80.3%, and the positive predictive value (PPV) was 92.3%. The researchers discovered that lymphocytes, eosinophils, basophils, and neutrophils were all strongly linked to COVID-19 diagnosis (Tordjman et al. 2020).

For Predicting

Huang et al. created a model to predict COVID-19 patients' progression to severe symptoms. The study analyzed data from 125 COVID-19 patients from Guangzhou's Eighth People's Hospital (93 with mild disease and 32 with severe disease). The model was created with the machine logistic regression, which has an AUC of 94.4%, a sensitivity of 94.1%, and a specificity of 90.2% (Huang et al. 2020).

Xie et al. created a model to predict COVID-19 patient mortality. A total of 444 patients from two hospitals were included in the investigation (Tongji Hospital and Jinyintan Hospital, Wuhan, China). The prediction model was built using the LR algorithm, which scored $c = 089$ and $c = 098$ for internal and external validation, respectively. Age, lymphocyte count, LDH, and SpO_2 were identified as four independent predictors for predicting death in the model. However, the authors noted that a few parameters, such as D-dimer and organ-specific damage markers, were not accessible in the study's cohorts and so were not included in the model (including cTnI, ALT, and BUN). This could have influenced the selection of prognostic criteria (Xie et al. 2020).

Zhao et al. created a risk-score model to predict ICU admission and mortality. The study examined data from 641 laboratory-confirmed COVID-19 patients from Stony Brook University Hospital in the United States (195 admitted to the ICU, 82 expired). Symptoms, comorbidities, demographics, laboratory data, vital signs, and imaging findings were all compared to those of non-critical COVID-19 patients in order to determine the most important characteristics predicting the two outcomes. On the testing dataset, the study used LR to obtain good accuracy, with an AUC of 0.83 for death prediction and 0.74 for ICU admission prediction. Elevated LDH, procalcitonin, and lowered SpO_2 were revealed to be the most common top predictors of death and ICU admission in the study. Furthermore, while a low lymphocyte count and a history of smoking were among the top predictors of ICU admission, they were not linked to an increased risk of death in this analysis. Cardiopulmonary markers and an elevated heart rate, on the other hand, were among the top predictors of death in COVID-19 patients, while ICU admission was not (Zhao et al. 2020).

In COVID-19 patients, Zhou et al. built a model to predict the severity of infection. The study analyzed data from 377 patients from Wuhan's Central Hospital (172 severe, 106 non-severe). The model was developed using the LR algorithm, and it had an AUC of 87.9%, a specificity of 73.7%, and a sensitivity of 88.6%. The findings indicated that three independent characteristics were linked to severity in COVID-19 patients: age, CRP, and D-dimer. Furthermore, the N/L*CRP*D-dimer product was discovered to be a significant predictor of illness severity (Zhou et al., 2020).

In the study by Onyijen et al., titled 'Prediction of Deaths from COVID-19 in Nigeria Using Various Machine Learning Algorithms,' a Linear Regression Model with an R^2 value of 1 successfully predicted total deaths from total cases, indicating that the evaluated variables were relevant. Multiple Linear Regression also provided a decent prediction, with an R^2 of 0.99 and an RMSE of 0.136. In determining the neural network model's performance, MAE and MSE of 0.0264 and 0.0013 respectively were reported. The 37,225 data collected comprised 13 variables. The data were prepared and the seven most important attributes were selected, namely total cases, total discharged, new cases, daily discharged, new deaths, aged 65 older and aged 70 older (Onyijen et al. 2021).

Support Vector Machine (SVM)

For Diagnosing

Soares et al. developed a machine learning algorithm to predict the diagnosis of COVID-19. The research looked at data from 599 Brazilian patients (81 positives). The model was trained using a combination of three strategies: SVM, SVM, and SVM. The sensitivity of the model is 63.98%, the specificity is 92.16%, and the NPV is 95.29% (Soares et al. 2020).

For Predicting

Booth et al. used Support Vector Machines (SVM) to predict death in COVID-19-positive patients using only a multiplex of serum biomarkers available from most clinical chemistry laboratories. The study utilized medical unit data from the University of Texas, including 398 patients (355 COVID-19 survivors and 43 non-survivors), to predict death up to 48 hours in advance. The highest-weighted laboratory values were chosen from the 26 measures initially collected: CRP, BUN, serum calcium, serum albumin, and lactic acid. For predicting patient death, the SVM model has 91% sensitivity and 91% specificity (AUC 0.93). When the whole dataset was analyzed, the study found that CRP, lactic acid, and serum calcium had the most significant impact and contributed the most to model outcomes (Booth et al. 2021).

Sun et al. created a model to predict COVID-19 patients' severe symptoms. A total of 336 patients were studied from the Shanghai Public Health Clinical Center. The prediction was built using the SVM model. The model chose four factors out of 220 clinical and laboratory features that made the most difference: age, GSH, CD3 ratio, and total protein. The AUC of the model was 97.57% (Sun et al. 2020).

Yao et al. developed a model to predict the severity of COVID-19 using data from blood or urine tests. The study included 137 patients from the Tongji hospital, Huazhong out of which 75 of them were critically ill. SVM was used to create the severity detection model, which had an accuracy of 81.48%. Age, blood test values, urine test values and pH were the top-ranking features recognized by the model (Yao et al. 2020).

Zhao et al. developed a model that can predict severity in patients with intermediate COVID-19. Univariate and multivariate LR models were used, six important features were chosen from a total of 22. The SVM technique was used to create the prediction model, which had an accuracy of 91.38%, a sensitivity of 0.90, and a specificity of 0.94. IL-6, high-sensitivity cardiac troponin I (cTnI), procalcitonin, hsCRP, chest distress, and calcium were the top six features for predicting severity (Zhao et al. 2021).

Tables 1.3 and 1.4 show various machine learning algorithms that have used for management of COVID-19 disease prediction and diagnosis respectively. For prediction, demographics and clinical datasets were mainly used and most researchers reviewed recorded prediction accuracy above 90%. On the other hand, clinical and demographics datasets as well as laboratory analysis data were the basis for the machine learning approach to diagnosing COVID-19. Table 1.4 shows various machine learning model for COVID-19 diagnosis and the respective accuracies.

TABLE 1.3
Recent Applications of Various Machine Learning Algorithms for COVID-19 Prediction and Management

S/N	ML Model	Data Type	Sample Size	Model Performance	Country	Reference
1	LR, ANN	Demographics, Clinical	37, 225 cases and 7 attributes	R2 = 99%RMSE = 0.136	Nigeria	Onyijen et al. (2021)
2	XGBoost	Clinical.	Blood samples of 75 features with a sample size of 485.	Accuracy: 90%	Wuhan, China	Yan et al. (2020)
3	XGBoost	Clinical, Demographics	3927 patients	AUC ranged between 92% and 81% using three validation cohorts.	USA	Bertsimas et al. (2020)
4	XGBoost	Clinical	4098 patients	AUCROC of 0.89 at 3 days, 0.85 at 5 and 7 days, and 0.84 at 10 days	New York, USA	Vaid et al. (2020)
5	RF	Demographics, Clinical	Total 253 samples from 169 patients.	Accuracy: 95.95%Specificity: 96.95%	China	Wu, Zhang, et al. (2020)
6	LR	Clinical	641 patients	AUC of 82%	USA	Zhao et al. (2020)
7	LR	Clinical	125 patients (32 severe)	AUC of 94.4%, sensitivity of 94.1%, specificity of 90.2%.	Guangzhou, China	Huang et al. (2020)
8	LR	Demographics, Clinical	444 patients	($c = 0{\cdot}89$) and ($c = 0{\cdot}98$) forinternal and external validation	Wuhan, China	Xie et al. (2020)
9	LR		377 patients	AUC 87.9%, specificity of 73.7%, sensitivity of 88.6%	Wuhan, China	Zhou et al. (2020)
10	SVM	Demographics, Clinical	137 COVID-19 patients	0.815 accuracy		Yao et al. (2020)
11	SVM	Clinical	303 patients	Accuracy of 91.38%, sensitivity of 0.90, specificity of 0.94	China	Zhao et al. (2021)
12	SVM	Clinical	398 patients	91% sensitivity and 91% specificity (AUC 0.93)	USA	Booth et al. (2021)
13	SVM	Clinical, Demographics, laboratory features	336 COVID-19 patients with 26 critical cases.	Accuracy: 77.5%, AUC: s99%, Specificity: 78.4%	China	Sun et al. (2020)
14	Autoencoder LR, RF, SVM, one-class SVM,	Clinical	Two datasets: A) 28,958PatientsB) 1448 patients	Autoencoder model achieved around 73% AUC, and 97% accuracy.	France	Li et al. (2020)
15	LR, SVM, GBDT, and NN	Demographics, Clinical	2520 COVID-19 patients	AUC ranging from 91.86% to 97.62%	China	Gao et al. (2020)

TABLE 1.4
Recent Applications of Various Machine Learning Methods for Diagnosis in COVID-19 Management

S/N	ML Model	Data Type	Sample Size	Model Performance	Country	Reference
1	XGBoost	Clinical	413 patients	Sensitivity of 92.5% and specificity of 97.9%	China	Li et al. (2020)
2	XGBoost	Clinical. Demographics	5333 patients (160 positive)	AUC of 97%, sensitivity of 81.9%, specificity of 97.9%	Slovenia	Kukar et al. (2021)
3	RF	Clinical blood test data	253 samples from 169 suspected patients	AUC of 99.26%, a sensitivity of 100%, and a specificity of 94.44%	China	Wu, Zhang, et al. (2020)
4	RF	Clinical	1528 patients (65 positive)	Accuracy of 81%, AUC of 0.74, sensitivity of 60%, and specificity of 82%		Tschoellitsch et al. (2021)
5	LR	Demographics, Clinical	43,752 surveys	AUC of 0.737	Israel	Shoer et al. (2021)
6	Binary LR	Clinical	400 patients (258 positive)	AUC of 88.9%, a sensitivity of 80.3%, and positive predictive value (PPV) of 92.3%.	France	Tordjman et al. (2020)
7	DT, ET, KNN, LR, NB, SVM,TWRF, and (RF selected)	Demographics, Clinical	279 patients (177 positive)	AUC of 84%, accuracy of 82%, sensitivity of 92%, PPV of 83%, and specificity of 65%	Italy	Brinati et al. (2020)
8	SVM	Clinical	599 patients (81 positive)	Specificity of 92.16%, NPV of 95.29%, and sensitivity of 63.98%	Brazil	Soares et al. (2020)
9	LR	Clinical, Demographics, laboratory features	2777 patients (368 PCR positive)	C-statistic of 78%, sensitivity of 86–93%, and specificity of 35–55%	USA	Joshi et al. (2020)
10	RF, NB, LR, SVM, and k- KNN	Clinical and Demographics	1624 patients (52% positive)	AUC 83% to 90%	Italy	Cabitza et al. (2020)
11	RF, LR, SVM, XGBoost, ADABoost	Clinical	1455 records (182 positive)	AUC of 91%, sensitivity of 93%, specificity of 64%	USA	Goodman-Meza et al. (2020)
12	RF, extra trees and LRXGBoost	Clinical blood test data	5644 patients (559positive)	AUC of 99.38%, sensitivity of 98.72% and specificity of 99.99%	Brazil	Aljame et al. (2020)
13	LR, DT, ADABoost, and Lasso	Demographic and Clinical blood test data	132 patients (26 positive)	AUC of 84.1% recall of 1.000, specificity of 0.727, and precision of 0.400	China	Feng et al. (2020)

RECENT RESEARCH ON MACHINE LEARNING APPROACH FOR COVID-19 PREDICTION AND DIAGNOSIS

Figure 1.7 shows the relationship between various machine learning algorithms and COVID-19 applications such as diagnosis and prediction. The figure presents database used by some the studies captured in this review. Patients database is a database of hospitals where patient data are stored. The figure further showed that these datasets can be clinical datasets or laboratory analysis data as seen in the oval shapes. This simply means that the datasets collected from the listed databases can either be clinical and laboratory. The data collected from the database are plugged into suitable machine learning algorithms such as; Random Forest, Logistics Regression, Support Vector Machine, XGBoost, ADABoost, Decision Tree, Neural Networks etc. for diagnosis and prediction depending on the type of data and objective of the researcher.

Some studies leveraged on several machine learning approaches for the predicting and diagnosing of patients with COVID-19. Some the selected studies are discussed as follows;

DIAGNOSING

Cabitza et al. developed algorithms for detecting COVID-19 positive patients using RF, NB, LR, SVM, and KNN. For this study, 1624 data of patients admitted to an hospital in Italy were trained using three different datasets. The entire data collection

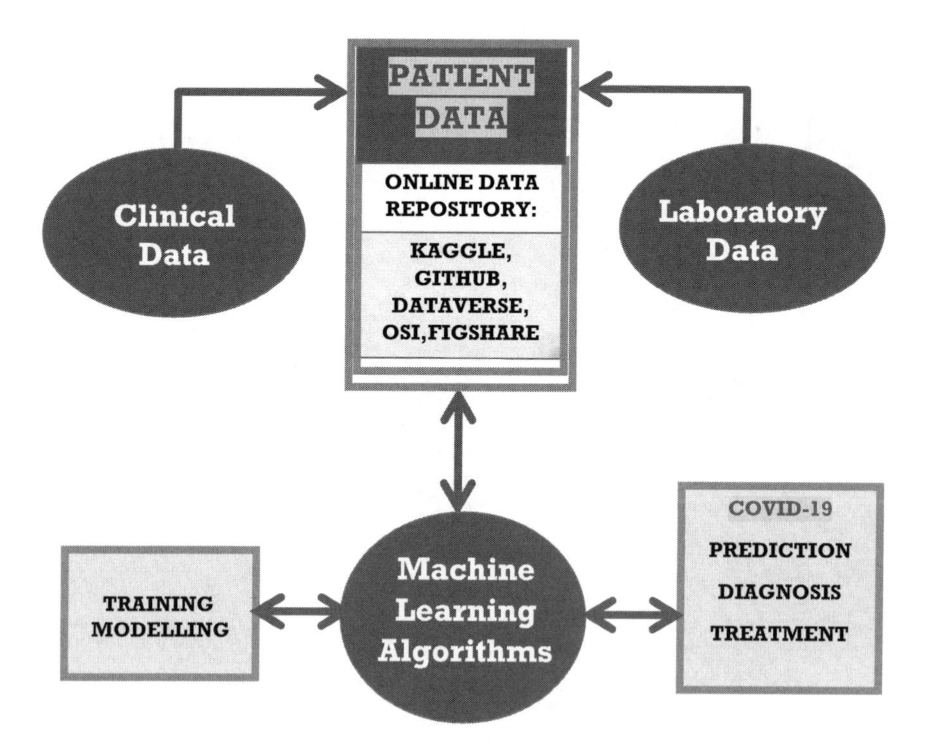

FIGURE 1.7 Relationship between machine learning algorithms and COVID-19 applications.

had 72 features that described various aspects of patient records, such as CBC, biochemical, coagulation, hemogas analysis, and CO-oxymetry values; age; sex; and particular symptoms at assessment. The other two sub-data sets consisted of 32 and 21 features, respectively. The models' AUCs ranged from 0.83 to 0.90. With AUCs ranging from 0.75 to 0.78 and specificity values ranging from 0.92 to 0.96, the internal–external validation yielded positive findings. The most important features, according to the study, were LDH, AST, CRP, and calcium. Age was also discovered to be a strong factor. WBC, fibrinogen, and cross-linked fibrin degradation products (XDPs) were other important features (Cabitza et al. 2020).

Goodman-Meza et al. constructed an ensemble of seven standard ML algorithms to diagnose COVID-19. The authors used RF, LR, SVM, multilayer perceptron, stochastic gradient descent, XGBoost, and Adaboost. Data from the UCLA Health System's computerized medical records were used in the study (Los Angeles, CA, USA). In total, 1455 records were found, with 1273 cases being negative and 182 being positive. The demographic and laboratory characteristics of each case were used to describe it. Experimentation demonstrated that including inflammatory markers in the model could improve the algorithm's prediction. CRP and LDH were two of the most critical characteristics for diagnosing COVID-19. Despite having a high sensitivity value, the proposed model had a high proportion of false positives (low PPV value). The model's accuracy in detecting negative cases was 64% (Goodman-Meza et al. 2020).

Routine blood tests were also employed by Aljame et al. to predict COVID-19. The researchers examined a publicly available dataset from Brazil, which included 5644 patients but only 559 COVID-19 positives. According to the model results, only 18 features were included, and they were sorted from the most weighted to the least weighted. The model started with a first level that included additional trees, RF, and LR, and then used the predictions from that level as input features for a second level that used XGBoost to improve performance. On the same dataset, the resultant performance was better than that of other models. The model had a 99.88% overall accuracy, a 99.38% AUC, a 98.72% sensitivity, and a 99.99% specificity (Aljame et al. 2020).

Feng et al. created a model for detecting COVID-19 patients early. The Chinese People's Liberation Army General Hospital in Beijing provided a dataset of 132 patients (26 positive). The study compared the performance of several ML approaches and discovered that LR with the least absolute shrinkage and selection operator (LASSO) performed best in the external validation and testing sets, with AUCs of 0.938 and 0.841, recall of 1.000 and 1.000, and specificity of 0.778 and 0.727, respectively, in the external validation and testing sets. The model revealed that age, interleukin-6 (IL-6), systolic blood pressure, monocyte ratio, and fever classification were the most important factors in predicting diagnosis (Feng et al. 2020).

Brinati et al. created a model to predict the diagnosis of COVID-19. The study analyzed data from 279 individuals at the IRCCS Ospedale San Raffaele in Italy. DT, ET, KNN, RF, LR, NB, SVM, and trees weighting RF were all used in the study. With an AUC of 84%, accuracy of 82%, sensitivity of 92%, and specificity of 65%, the RF classifier was found to have the best performance. AST, lymphocytes, LDH, WBC, eosinophils, alanine transaminase (ALT), and age were the most critical factors in predicting COVID-19 diagnosis (Brinati et al. 2020).

Prediction

Given a patient's underlying health problems, age, sex, and other characteristics, Li et al. constructed a model to estimate the mortality risk of COVID-19 patients. The GitHub dataset, which had 28,958 instances (530 deaths) after processing, and the Wolfram dataset, which included 1448 records, were both used (123 deaths). The Wolfram dataset produced better results than the Github dataset, which lacked exact information and largely provided broad information on each case, limiting the predictive capacity of models. LR, RF, SVM, one-class SVM, isolation forest, local outlier factor, and autoencoder were among the ML and data learning techniques used. The greatest results were from the autoencoder model, which had a 73% AUC, 97% accuracy, 97% specificity, and 40% sensitivity. The findings revealed that having a chronic disease or symptoms of the gastrointestinal, kidney, cardiac, or respiratory systems was significantly linked to patient death (Li et al. 2020).

Gao et al. created a COVID-19 death prediction model that incorporates clinical data from patients' EHRs on admission to allow for accurate and timely mortality risk classification of patients up to 20 days in advance. The study included 2520 COVID-19 patients with known outcomes (discharge or death) from two Tongji Medical College affiliated hospitals. The model was created using four machine learning techniques: LR, SVM, GBDT, and NN. The study chose 14 features out of 34 for modeling. An internal validation cohort and two external validation cohorts were used to validate the model, with AUCs of 96.21%, 97.60%, and 92.46%, respectively (Gao et al. 2020).

CONCLUSION

The relevance of the machine learning techniques in the prognosis and management of the COVID-19 pandemic is presented in this chapter. The global perspective to this important topic is given from the recent studies using machine learning techniques to address various aspects of the COVID-19 pandemic management. Machine learning's involvement in solving the issues of COVID-19 detection, diagnosis, and management, as well as categorization, medication discovery, and prognosis, is discussed. COVID-19 illness and its variants are discussed, as well as the mode of transmission. The latest breakthroughs in machine learning algorithms for COVID-19 disease prognosis, prediction, diagnosis, therapies, and management are reviewed. Most models have accuracy rates of around 90%, indicating the importance of machine learning techniques as viable tools for managing the COVID-19 pandemic and a plausible field of research for tackling the threat of the COVID-19 pandemic on the world's socioeconomic well-being.

REFERENCES

Aljame, M., Ahmad, I., Imtiaz, A., and Mohammed, A., 2020. Ensemble learning model for diagnosing COVID-19 from routine blood tests. *Information in Medicine*, 21 (January), 2–11.

Anani, O.A., Adetunji, C.O., Olugbemi, O.T., Hefft, D.I., Wilson, N., and Olayinka, A.S., 2022. IoT-based monitoring system for freshwater fish farming: Analysis and design. In *AI, Edge and IoT-based Smart Agriculture* (pp. 505–515). Elsevier. https://doi.org/10.1016/B978-0-12-823694-9.00026-8

Bertsimas, D., Lukin, G., Mingardi, L., Nohadani, O., Orfanoudaki, A., Stellato, B., Wiberg, H., Gonzalez-Garcia, S., Parra-Calderón, C.L., Robinson, K., Schneider, M., Stein, B., Estirado, A., Beccara, L.A., Canino, R., Bello, M.D., Pezzetti, F., Pan, A., Akinosoglou, K., Antoniadou, A., Argyraki, K., Dalekos, G.N., Gaga, M., Gatselis, N.K., Gogos, C., Kalomenidis, I., Kranidioti, E., Korompoki, E., Lourida, G., Margellou, E., Ntaios, G., Panagopoulos, P., Pefanis, A., Petrakis, V., Psarrakis, C., Sakka, V., Thomas, K., and Zervas, E., 2020. COVID-19 mortality risk assessment: An international multi-center study. *PLoS ONE*, 15 (12 December), 1–13.

Booth, A.L., Abels, E., and McCaffrey, P., 2021. Development of a prognostic model for mortality in COVID-19 infection using machine learning. *Modern Pathology*, 34 (3), 522–531.

Brinati, D., Campagner, A., Ferrari, D., Locatelli, M., Banfi, G., and Cabitza, F., 2020. Detection of COVID-19 infection from routine blood exams with machine learning: A feasibility study. *Journal of Medical Systems*, 44(8), 135.

Cabitza, F. Campagner, A., Ferrari, D., Di Resta, C., Ceriotti, D., Sabetta, E., Colombini, A., De Vecchi, E., Banfi, G., Locatelli, M., and Carobene, A., 2020. Development, evaluation, and validation of machine learning models for COVID-19 detection based on routine blood tests. *medRxiv*, 38 (August), 1–38.

Calisher, C.H., Childs, J.E., Field, H.E., Holmes, K. V., and Schountz, T., 2006. Bats: Important reservoir hosts of emerging viruses. *Clinical Microbiology Reviews*. https://journals.asm.org/doi/full/10.1128/cmr.00017-06

CDC, 2022a. Human coronavirus types, *CDC* [online]. Available from: www.cdc.gov/coronavirus/types.html [Accessed 24 Jan 2022].

CDC, 2022b. Delta variant: What we know about the science, *CDC* [online]. Available from: www.cdc.gov/coronavirus/2019-ncov/variants/delta-variant.html [Accessed 24 Jan 2022].

Cheng, V.C.C., Lau, S.K.P., Woo, P.C.Y., and Yuen, Y., 2007. Severe acute respiratory syndrome coronavirus as an agent of emerging and reemerging infection. *Clinical Microbiology Reviews*, 20 (4), 660–694.

Chowell, G., Abdirizak, F., Lee, S., Lee, J., Jung, E., Nishiura, H., and Viboud, C., 2015. Transmission characteristics of MERS and SARS in the healthcare setting: A comparative study. *BMC Medicine*, 13 (1), 210.

Feng, Z., Diao, B., Wang, R., Wang, G., Wang, C., Tan, Y., Liu, L., Wang, C., Liu, Y., Liu, Y., Yuan, Z., Ren, L., Wu, Y., and Chen, Y., 2020. The novel severe acute respiratory syndrome coronavirus 2 (SARS-CoV-2) directly decimates human spleens and lymph nodes. *medRxiv*, 2, 1–18.

Fontanet, A., Autran, B., Lina, B., Kieny, M.P., Karim, S.S.A., and Sridhar, D., 2021. SARS-CoV-2 variants and ending the COVID-19 pandemic. *The Lancet*, 397 (10278), 952–954.

Gao, Y., Cai, G.Y., Fang, W., Li, H.Y., Wang, S.Y., Chen, L., Yu, Y., Liu, D., Xu, S., Cui, P.F., Zeng, S.Q., Feng, X.X., Yu, R. Di, Wang, Y., Yuan, Y., Jiao, X.F., Chi, J.H., Liu, J.H., Li, R.Y., Zheng, X., Song, C.Y., Jin, N., Gong, W.J., Liu, X.Y., Huang, L., Tian, X., Li, L., Xing, H., Ma, D., Li, C.R., Ye, F., and Gao, Q.L., 2020. Machine learning based early warning system enables accurate mortality risk prediction for COVID-19. *Nature Communications*, 11 (1), 1–10.

Ghandour, A.J., Hammoud, H., and Al-Hajj, S., 2020. Analyzing factors associated with fatal road crashes: A machine learning approach. *International Journal of Environmental Research and Public Health*, 17 (11), 4111.

Gianfranco, F., Soddu, S., and Fadda, P., 2017. An accident prediction model for urban road networks. *Journal of Transportation Safety & Security*, 10 (4), 387–405. https://doi.org/10.1080/19439962.2016.1268659

Giovanetti, M., Benvenuto, D., Angeletti, S., and Ciccozzi, M., 2020. The first two cases of 2019-nCoV in Italy: Where they come from? *Journal of Medical Virology*, 92 (5), 518–521.

Giovanetti, M., Cella, E., Benedetti, F., Rife Magalis, B., Fonseca, V., Fabris, S., Campisi, G., Ciccozzi, A., Angeletti, S., Borsetti, A., Tambone, V., Sagnelli, C., Pascarella, S., Riva, A., Ceccarelli, G., Marcello, A., Azarian, T., Wilkinson, E., de Oliveira, T., Alcantara, L.C.J., Cauda, R., Caruso, A., Dean, N.E., Browne, C., Lourenco, J., Salemi, M., Zella, D., and Ciccozzi, M., 2021. SARS-CoV-2 shifting transmission dynamics and hidden reservoirs potentially limit efficacy of public health interventions in Italy. *Communications Biology*, 4 (1), 489.

Goodman-Meza, D., Rudas, A., Chiang, J.N., Adamson, P.C., Ebinger, J., Sun, N., Botting, P., Fulcher, J.A., Saab, F.G., Brook, R., Eskin, E., An, U., Kordi, M., Jew, B., Balliu, B., Chen, Z., Hill, B.L., Rahmani, E., Halperin, E., and Manuel, V., 2020. A machine learning algorithm to increase COVID-19 inpatient diagnostic capacity. *PLoS ONE*, 15 (9 September), 1–10.

Guan, W., Ni, Z., Hu, Y., Liang, W., Ou, C., He, J., Liu, L., Shan, H., Lei, C., Hui, D.S.C., Du, B., Li, L., Zeng, G., Yuen, K.-Y., Chen, R., Tang, C., Wang, T., Chen, P., Xiang, J., Li, S., Wang, J., Liang, Z., Peng, Y., Wei, L., Liu, Y., Hu, Y., Peng, P., Wang, J., Liu, J., Chen, Z., Li, G., Zheng, Z., Qiu, S., Luo, J., Ye, C., Zhu, S., and Zhong, N., 2020. Clinical characteristics of coronavirus disease 2019 in China. *New England Journal of Medicine*, 382 (18), 1708–1720.

Gutierrez-Osorio, C. and Pedraza, C., 2020. Modern data sources and techniques for analysis and forecast of road accidents: A review. *Journal of Traffic and Transportation Engineering (English Edition)*, 7 (4), 432–446.

Hu, B., Guo, H., Zhou, P., and Shi, Z.L., 2021. Characteristics of SARS-CoV-2 and COVID-19. *Nature Reviews Microbiology*, 19, 141–154

Huang, C., Wang, Y., Li, X., Ren, L., Zhao, J., Hu, Y., Zhang, L., Fan, G., Xu, J., Gu, X., Cheng, Z., Yu, T., Xia, J., Wei, Y., Wu, W., Xie, X., Yin, W., Li, H., Liu, M., Xiao, Y., Gao, H., Guo, L., Xie, J., Wang, G., Jiang, R., Gao, Z., Jin, Q., Wang, J., and Cao, B., 2020. Clinical features of patients infected with 2019 novel coronavirus in Wuhan, China. *The Lancet*, 395 (10223), 497–506.

Irving, A.T., Ahn, M., Goh, G., Anderson, D.E., and Wang, L.F., 2021. Lessons from the host defences of bats, a unique viral reservoir. *Nature*, 584, 363–370.

Iwasaki, Y., Takeuchi, I., Stanev, V., Kusne, A.G., Ishida, M., Kirihara, A., Ihara, K., Sawada, R., Terashima, K., Someya, H., Uchida, K., Ichi, S.E., and Yorozu, S. 2019. Machine-learning guided discovery of a new thermoelectric material. *Scientific Reports*, 9 (1), 2751.

Joshi, R.P., Pejaver, V., Hammarlund, N.E., Sung, H., Lee, S.K., Furmanchuk, A., Lee, H.Y., Scott, G., Gombar, S., Shah, N., Shen, S., Nassiri, A., Schneider, D., Ahmad, F.S., Liebovitz, D., Kho, A., Mooney, S., Pinsky, B.A., and Banaei, N., 2020. A predictive tool for identification of SARS-CoV-2 PCR-negative emergency department patients using routine test results. *Journal of Clinical Virology*, 129 (June), 104502.

Kang, C.K., Song, K.H., Choe, P.G., Park, W.B., Bang, J.H., Kim, E.S., Park, S.W., Kim, H. Bin, Kim, N.J., Cho, S. Il, Lee, J.K., and Oh, M.D., 2017. Clinical and epidemiologic characteristics of spreaders of middle east respiratory syndrome coronavirus during the 2015 outbreak in Korea. *Journal of Korean Medical Science*, 32 (5), 744–749.

Kukar, M., Gunčar, G., Vovko, T., Podnar, S., Černelč, P., Brvar, M., Zalaznik, M., Notar, M., Moškon, S., and Notar, M., 2021. COVID-19 diagnosis by routine blood tests using machine learning. *Scientific Reports*, 11 (1), 1–9.

Li, Y., Horowitz, M.A., Liu, J., Chew, A., Lan, H., Liu, Q., Sha, D., and Yang, C., 2020. Individual-Level Fatality Prediction of COVID-19 patients using AI Methods. *Frontiers in Public Health*, 8 (September), 1–12.

Liu, Y., Esan, O.C., Pan, Z., and An, L., 2021. Machine learning for advanced energy materials. *Energy and AI*, 3, 100049.

Morens, D.M., Breman, J.G., Calisher, C.H., Doherty, P.C., Hahn, B.H., Keusch, G.T., Kramer, L.D., Leduc, J.W., Monath, T.P., and Taubenberger, J.K., 2020. Perspective piece the origin of COVID-19 and why it matters. *Am. J. Trop. Med. Hyg*, 103 (3), 955–959.

Nwankwo, W., Adetunji, C.O., and Olayinka, A.S., 2022. IoT-driven bayesian learning: A case study of reducing road accidents of commercial vehicles on highways. In: S. Pal, D. De, and R. Buyya, eds. *Artificial intelligence-based internet of things systems*. Springer International Publishing, 391–418.

Olayinka, A.S., Adetunji, C.O., Nwankwo, W., Olugbemi, O.T., and Olayinka, T.C., 2022. A study on the application of bayesian learning and decision trees IoT-enabled system in postharvest storage. In: S. Pal, D. De, and R. Buyya, eds. *Artificial intelligence-based internet of things systems*. Springer International Publishing, 467–491.

Olayinka, A.S., Nwankwo, W., and Olayinka, T.C., 2020. Model based machine learning approach to predict thermoelectric figure of merit. *Archive of Science & Technology*, 1, 55–67.

Onyijen, O. H., Hamadani, A., Awojide, S., Ebhohimen, I.E., 2021. Prediction of deaths from COVID-19 in Nigeria using various machine learning algorithms. *SAU Sci-Technology Journal*, 6 (1), 109–117.

Ozili, P.K. and Arun, T., 2020. Spillover of COVID-19: Impact on the global economy. *SSRN Electronic Journal* (March 27, 2020). Available at SSRN: https://ssrn.com/abstract=3562570 or http://dx.doi.org/10.2139/ssrn.3562570

Platto, S., Wang, Y., Zhou, J., and Carafoli, E., 2021. History of the COVID-19 pandemic: Origin, explosion, worldwide spreading. *Biochemical and Biophysical Research Communications*, 538, 14–23.

Pollice, R., Dos Passos Gomes, G., Aldeghi, M., Hickman, R.J., Krenn, M., Lavigne, C., Lindner-D'Addario, M., Nigam, A., Ser, C.T., Yao, Z., and Aspuru-Guzik, A., 2021. Data-driven strategies for accelerated materials design. *Accounts of Chemical Research*, 54 (4), 849–860.

Punn, N.S., Sonbhadra, S.K., and Agarwal, S., 2020. COVID-19 epidemic analysis using machine learning and deep learning algorithms. *medRxiv*. https://doi.org/10.1101/2020.04.08.20057679

Raccuglia, P., Elbert, K.C., Adler, P.D.F., Falk, C., Wenny, M.B., Mollo, A., Zeller, M., Friedler, S.A., Schrier, J., and Norquist, A.J., 2016. Machine-learning-assisted materials discovery using failed experiments. *Nature*, 533 (7601), 73–76.

Rajkumar, A.R., Prabhakar, S., and Priyadharsini, A.M., 2020. Prediction of road accident severity using machine learning algorithm. *International Journal of Advanced Science and Technology*, 29 (6), 116–120.

Shereen, M.A., Khan, S., Kazmi, A., Bashir, N., and Siddique, R., 2020. COVID-19 infection: Emergence, transmission, and characteristics of human coronaviruses. *Journal of Advanced Research*, 24, 91–98.

Shoer, S., Karady, T., Keshet, A., Shilo, S., Rossman, H., Gavrieli, A., Meir, T., Lavon, A., Kolobkov, D., Kalka, I., Godneva, A., Cohen, O., Kariv, A., Hoch, O., Zer-Aviv, M., Castel, N., Sudre, C., Zohar, A.E., Irony, A., Spector, T., Geiger, B., Hizi, D., Shalev, V., Balicer, R., and Segal, E., 2021. A prediction model to prioritize individuals for a SARS-CoV-2 test built from national symptom surveys. *Med*, 2 (2), 196.e4–208.e4.

Soares, F. Villavicencio, A. Anzanello, M. J. Fogliatto, Idiart, F. S., and Stevenson, M. 2020. A novel high specificity COVID-19 screening method based on simple blood exams and artificial intelligence. *medRxiv*, 2–19. https://doi.org/10.1101/2020.04.10.20061036

Sun, L., Song, F., Shi, N., Liu, F., Li, S., Li, P., Zhang, W., Jiang, X., Zhang, Y., Sun, L., Chen, X., and Shi, Y., 2020. Combination of four clinical indicators predicts the severe/critical symptom of patients infected COVID-19. *Journal of Clinical Virology*, 128, 104431.

Tao, H., Wu, T., Aldeghi, M., Wu, T.C., Aspuru-Guzik, A., and Kumacheva, E., 2021. Nanoparticle synthesis assisted by machine learning. *Nature Reviews Materials*, 6, 701–716.

Tordjman, M., Mekki, A., Mali, R.D., Saab, I., Chassagnon, G., Guillo, E., Burns, R., Eshagh, D., Beaune, S., Madelin, G., Bessis, S., Feydy, A., Mihoubi, F., Doumenc, B., Mouthon, L., Carlier, R.Y., Drapé, J.L., and Revel, M.P., 2020. Pre-test probability for SARS-Cov-2-related infection score: The PARIS score. *PLoS ONE*, 15 (12 December), 1–14.

Townsend, J.P., Hassler, H.B., Wang, Z., Miura, S., Singh, J., Kumar, S., Ruddle, N.H., Galvani, A.P., and Dornburg, A., 2021. The durability of immunity against reinfection by SARS-CoV-2: a comparative evolutionary study. *The Lancet Microbe*, 2 (12), e666–e675.

Tschoellitsch, T., Dünser, M., Böck, C., Schwarzbauer, K., and Meier, J., 2021. Machine learning prediction of SARS-CoV-2 polymerase chain reaction results with routine blood tests. *Laboratory medicine*, 52 (2), 146–149.

UNICEF. 2022. What we know about the omicron variant. UNICEF [online]. Available from: https://www.unicef.org/coronavirus/what-we-know-about-omicron-variant [Accessed 24 Jan 2022].

Vaid, A., Somani, S., Russak, A.J., de Freitas, J.K., Chaudhry, F.F., Paranjpe, I., Johnson, K.W., Lee, S.J., Miotto, R., Richter, F., Zhao, S., Beckmann, N.D., Naik, N., Kia, A., Timsina, P., Lala, A., Paranjpe, M., Golden, E., Danieletto, M., Singh, M., Meyer, D., O'Reilly, P.F., Huckins, L., Kovatch, P., Finkelstein, J., Freeman, R.M., Argulian, E., Kasarskis, A., Percha, B., Aberg, J.A., Bagiella, E., Horowitz, C.R., Murphy, B., Nestler, E.J., Schadt, E.E., Cho, J.H., Cordon-Cardo, C., Fuster, V., Charney, D.S., Reich, D.L., Bottinger, E.P., Levin, M.A., Narula, J., Fayad, Z.A., Just, A.C., Charney, A.W., Nadkarni, G.N., and Glicksberg, B.S., 2020. Machine learning to predict mortality and critical events in a cohort of patients with COVID-19 in New York City: Model development and validation. *Journal of Medical Internet Research*, 22 (11), 1–19.

WHO. 2022. WHO coronavirus (COVID-19) dashboard with vaccination data [online]. Available from: https://covid19.who.int/ [Accessed 19 January 2022].

Wu, F., Zhao, S., Yu, B., Chen, Y.M., Wang, W., Song, Z.G., Hu, Y., Tao, Z.W., Tian, J.H., Pei, Y.Y., Yuan, M.L., Zhang, Y.L., Dai, F.H., Liu, Y., Wang, Q.M., Zheng, J.J., Xu, L., Holmes, E.C., and Zhang, Y.Z., 2020. A new coronavirus associated with human respiratory disease in China. *Nature*, 579 (7798), 265–269.

Wu, J., Zhang, P., Zhang, L., Meng, W., Li, J., Tong, C., Li, Y., Cai, J., Yang, Z., Zhu, J., Zhao, M., Huang, H., Xie, X., and Li, S., 2020. Rapid and accurate identification of COVID-19 infection through machine learning based on clinical available blood test results. *medRxiv*. https://doi.org/10.1101/2020.04.02.20051136

Xie, J., Hungerford, D., Chen, H., Abrams, S.T., and Li, S., 2020. Development and external validation of a prognostic multivariable model on admission for hospitalized patients with COVID-19. 1–22.

Yan, L., Zhang, H.-T., Goncalves, J., Xiao, Y., Wang, M., Guo, Y., Sun, C., Tang, X., Jing, L., Zhang, M., Huang, X., Xiao, Y., Cao, H., Chen, Y., Ren, T., Wang, F., Xiao, Y., Huang, S., Tan, X., Huang, N., Jiao, B., Cheng, C., Zhang, Y., Luo, A., Mombaerts, L., Jin, J., Cao, Z., Li, S., Xu, H., and Yuan, Y., 2020. An interpretable mortality prediction model for COVID-19 patients. *Nature Machine Intelligence*, 2 (5), 283–288.

Yao, H., Zhang, N., Zhang, R., Duan, M., Xie, T., Pan, J., Peng, E., Huang, J., Zhang, Y., Xu, X., Xu, H., Zhou, F., and Wang, G., 2020. Severity detection for the coronavirus disease 2019 (COVID-19) patients using a machine learning model based on the blood and urine tests. *Frontiers in Cell and Developmental Biology*, 8 (July), 1–10.

Yin, Y. and Wunderink, R.G., 2018. MERS, SARS and other coronaviruses as causes of pneumonia. *Respirology*, 23 (2), 130–137.

Zhao, C., Bai, Y., Wang, C., Zhong, Y., Lu, N., Tian, L., Cai, F., and Jin, R., 2021. Risk factors related to the severity of COVID-19 in Wuhan. *International Journal of Medical Sciences*, 18 (1), 120–127.

Zhao, Z., Chen, A., Hou, W., Graham, J.M., Li, H., Richman, P.S., Thode, H.C., Singer, A.J., and Duong, T.Q., 2020. Prediction model and risk scores of ICU admission and mortality in COVID-19. *PLoS ONE*, 15 (7 July), 1–14.

Zhou, P., Yang, X. Lou, Wang, X.G., Hu, B., Zhang, L., Zhang, W., Si, H.R., Zhu, Y., Li, B., Huang, C.L., Chen, H.D., Chen, J., Luo, Y., Guo, H., Jiang, R. Di, Liu, M.Q., Chen, Y., Shen, X.R., Wang, X., Zheng, X.S., Zhao, K., Chen, Q.J., Deng, F., Liu, L.L., Yan, B., Zhan, F.X., Wang, Y.Y., Xiao, G.F., and Shi, Z.L., 2020. A pneumonia outbreak associated with a new coronavirus of probable bat origin. *Nature*, 579 (7798), 270–273.

Zhu, N., Zhang, D., Wang, W., Li, X., Yang, B., Song, J., Zhao, X., Huang, B., Shi, W., Lu, R., Niu, P., Zhan, F., Ma, X., Wang, D., Xu, W., Wu, G., Gao, G.F., and Tan, W., 2020. A novel coronavirus from patients with pneumonia in China, 2019. *New England Journal of Medicine*, 38 (8), 727–733.

2 Application of Artificial Intelligence in the Handling of COVID-19

Igiku Victory, Charles Oluwaseun Adetunji, and Akinola Samson Olayinka
Edo State University, Uzairue, Nigeria

Olugbemi T. Olaniyan
Rhema University, Aba, Nigeria

Juliana Bunmi Adetunji
Osun State University, Osogbo, Nigeria

Tosin Comfort Olayinka
Wellspring University, Benin City, Nigeria

Aishatu Idris Habib
Edo State University, Uzairue, Nigeria

Olorunsola Adeyomoye and Oluwafemi Adebayo Oyewole
Federal University of Technology, Minna, Nigeria

K. I. T. Eniola
Joseph Ayo Babalola University, Ikeji Arakeji, Nigeria

2.1 INTRODUCTION

COVID-19 is a disease with tremendous global impact. Originating from China at the end of 2019, there have been a total of 526,932,367 confirmed cases with a death toll of 6,299,394, and it has been reported all over the world in 227 countries. The World Health Organization (WHO), announced the disease was a global pandemic on March 11, 2020. The disease is caused by the severe acute respiratory syndrome coronavirus 2 (SARS-CoV-2).

DOI: 10.1201/9781003309468-2

The spread of the SARS-CoV-2 surpassed that of both severe acute respiratory syndrome and the Middle Eastern respiratory syndrome in terms of the spread of the virus and the rate of infection (Hu et al. 2020). SARS-COV-2 has a genomic sequence that is 79% similar to SARS-CoV, 50% similar to MERS-CoV (Lu 2019), and 96.2% similar to bat CoV-RaTG13 (Zhou et al., 2020a,b). Due to this similarity, Ludwig and Zenbock (2020), suggested that the virus originated from bats and may have spread from bats to other animal and humans. According to WHO, the disease spreads from person to person through contact transmission, saliva, droplets, or nose discharge. Also, the virus has been reported in anal swabs and blood of infected patients (Zhang et al. 2020).

The symptoms of the disease include fever, fatigue, dyspnea, cough, loss of smell, and in severe cases pneumonia, acute respiratory disease syndrome, and multiple organ failure (Zen et al. 2020). Artificial intelligence is a viable tool in managing the current human challenges.

Industrial, medical, and science research has effectively utilized advanced artificial intelligence technology in handling COVID-19 with remarkable success within the short time since the outbreak of the disease (Alimadai et al. 2020). Digital pathology and artificial intelligence, according to Browning et al. (2021), will be very useful in aiding clinical and academic cellular pathology.

Shortly after the outbreak of the COVID-19, digital healthcare services such as full-scale monitoring technologies, therapeutic decision-making methods, telehealth, creative diagnostic were extensively adopted to respond to the public health emergency. The white house in partnership with research institutes and technology companies on 16 March, 2020, urged artificial intelligence scientist to come up with new text and data mining methods that can aid COVID-19 research. The success of artificial intelligence depends on big data. Artificial intelligence and machine language, were successfully applied in COVID-19 due to the increase in the availability of big data (Mhlanga, 2022). Due to this abundance of data, it has become possible to predict and track the spread of the disease using artificial intelligence technologies (Chen and See, 2020).

Artificial intelligence analyzes COVID-19 big data and social media data to develop different frameworks that will be able to predict with accuracy the spread of the disease (Alimadadi et al. 2020; Adetunji et al., 2020; Oyedara et al., 2022; Adetunji et al., 2022a, b, c, d, e, f, g, h; Olaniyan et al., 2022a, b).

Machine learning algorithms may be used to evaluate COVID-19 big data in order to anticipate the dynamics of viral transmission, enhance diagnostic speed, develop new treatment methods, and identify persons at risk based on genetic predisposition and physiological characteristics (Alimadadi et al. 2020).

Artificial intelligence, according to the World Health Organization (WHO), might be a viable means of combating the epidemic. Artificial intelligence technology has been used in public health to forecast the spread of the infection and determine the effect of control measures, diagnose, and monitor the disease and in epidemiology study (Asada et al. 2021). This technology can speed up the diagnosis and treatment of patients with COVID-19 (Jamshidi et al., 2020).

Early warning and alerts, tracking and prediction, data dashboard, diagnosis and prognosis, treatments and cures, and social control are six areas where artificial intelligence contributes to the management of COVID-19, according to Naudé (2020b).

Islam et al. (2021) stated that artificial intelligence was essential in the diagnosis, classification, detection, prediction of disease severity and mortality risk of the disease. Bullock et al. (2020), identified the application of artificial intelligence in the handling of COVID-19 at three different levels, namely molecular level, clinical level, and societal level. At the molecular level, artificial intelligence technology such as AlphaFold can be useful in the prediction of the SARS-CoV-2 viral protein, identification of existing drug candidates that can be repurposed to fight the virus, find and propose prospective novel compounds for drug development, identify potential vaccine targets, develop new diagnostic methods, and assist researchers and healthcare providers in understanding viral infectivity and disease severity. At the clinical level, artificial intelligence technologies can be used in the diagnosis of COVID-19 through the analysis of medical images, provide alternative ways to track the spread of the disease and predict the possible outcome of COVID-19 patients based on data inputs such as electronic health records. Finally at the societal level, artificial intelligence can be used in different areas of epidemiological study involving the modelling of empirical data, such as forecasting of cases given different choices of public policy (Bullock et al. 2020). The major scope of artificial intelligence in handling COVID-19 covered in this chapter includes the application of artificial intelligence in the diagnosis of COVID-19, drug discovery, and epidemiology and social control.

2.2 APPLICATION OF ARTIFICIAL INTELLIGENCE IN THE DIAGNOSIS OF COVID-19

Early diagnosis of COVID-19 is of great importance in the management of the pandemic as it enables healthcare workers and the government authorities to break the disease cycle and therefore halt the spread of the disease. In order to achieve optimal outcomes efficient diagnosis of the clinical type of COVID-19 is necessary (Mohamadou et al., 2020). The conventional COVID-19 diagnosis method is the RT-PCR detects the presence of the COVID-19 virus genetic material using reverse transcription polymerase chain reaction. This method of diagnosis on the other hand, has been identified to be costly, and the results take longer. Antigen tests using IGM/IGG antibodies have been employed to detect the virus; however, their sensitivity and specificity are limited (Mohamadou et al., 2020).

Most artificial intelligence assisted diagnostic procedures utilize medical imaging such as X-ray and computed tomography (CT), radiology, and predictive analysis in diagnostic procedures. A study by Wynants et al. (2020), identified 118 diagnostic models for detecting COVID-19, of which 63.6% of the model was based on medical imaging and approximately 8.5% of the model was used to diagnose disease severity. Artificial intelligence also presents the opportunity to acquire images and scan patients remotely, which decreases contact between health workers and COVID-19 patients and in turn reduces the risk of infection. When compared to the traditional detection approach, artificial intelligence-based COVID-19 detection is safe, accurate, and quick (Swayamsiddha et al., 2021).

Artificial intelligence algorithms can help in the diagnosis of COVID-19 by analyzing clinical symptoms, exposure history, and laboratory testing (Swayamsiddha et al., 2021). X-rays and CT imaging are important medical imaging tools in the fight against COVID-19; artificial intelligence empowers these tools further by improving the imaging power of these tools and increases the efficiency of medical specialists (Shi et al. 2020).

Infervision, an artificial intelligence medical startup company in China, launched a medical imaging artificial intelligence framework, inferRead, which uses both deep learning technology and neural network models to diagnose COVID-19 cases through the identification of specific lung features (Bragazzi et al., 2020). According to the Sinai Health System (2020), researchers in mount Sinai were the first to use artificial intelligence to diagnose patients with COVID-19 in the United States. The researchers used artificial intelligence algorithms combined with CT imaging and clinical data such as patient symptoms, exposure history, and laboratory test to diagnose patients with COVID-19. The result of the study showed that the artificial intelligence model had an area under the curve (AUC) of 0.92 and had equal sensitivity compared to a radiologist (Mei et al. 2020). The model can be an effective alternative for the diagnosis of COVID-19.

Researchers have proposed various artificial intelligence models which could be useful for the diagnosis of COVID-19. McCall (2020) reported the application of artificial intelligence technology in analyzing large number of CT images to speed up the diagnosis of the disease. Polsinelli et al. (2020) developed a convolutional neural network (CNN) model based on the SqueezeNet model for the detection of COVID-19 from CT images. Similarly, Abraham and Nair (2020), reported the efficacy of several pre-trained CNN models namely, Darknet-53, MobilenetV3, Xception, and Shufflenet for the detection of COVID-19 from X-ray images using two datasets. The study showed that in an area under the curve of 0.963, accuracy of 91.16% was achieved in the first datasets consisting of 453 COVID-19 images and 495 non COVID-19 images. Furthermore, an AUC of 0.911 and accuracy of 97.44% was achieved in the second dataset consisting of 71 COVID-19 images and 7 non-COVID-19 images. Ouchicha et al. (2020) proposed the use of a deep convolutional neural network (DCNN) model, which could differentiate COVID-19 infection from normal and other pneumonia cases using X-ray images. Gupta et al. (2020), proposed the use of a DCNN called InstaCovNet-19, which combines several pre - trained models such as ResNet101, Xception, InceptionV3, MobileNet and NAsNet to detect COVID-19 from different chest x-ray images. Li et al. (2020), deduced a correlation between males and higher levels serum lymphocytes and neutrophil levels using a machine language model called XGBoost, to reanalyze COVID-19 data from 151 published studies. The study classified COVID-19 patients into three, according to immune serum levels, gender and symptoms. The model used in this study achieved a sensitivity of 92.5% and specificity of 97.9%. in a study by Chen et al. (2021), the author, reported the use of a random forest model based on machine language, which classified COVID-19 clinical types with a predictive accuracy of more than 90%.

In another study, Liang et al. (2022) proposed an artificial intelligence framework called CoviDet, which is based on deep learning (DL) and federated learning (FL) for COVID-19 diagnosis, monitoring and prediction of patients clinical course based on CT scans alone. The proposed model showed an area under the curve (AUC) of 0.98% and had high consistency in the diagnosis of COVID-19 using CT scans. The model can diagnose COVID-19 using CT scans with or without the availability of clinical data.

According to Chen et al. (2021), the steps involved in the analysis of CT image by artificial intelligence includes: (1) the segmentation of region of interest, (2) extraction of lung tissue features, (3) detection of infected region, (4) classification of the disease. Lung and region of interest segmentation is the basic step in the artificial intelligence assisted medical image inspection. This process shows the regions of interest in the CT images of the lungs such as the lungs, lung lobes, bronchopulmonary segments, and infected regions. Several deep learning models have been used for the segmentation of CT images such as U-Net, V-Net, and VB-Net (Chen and See, 2020; Xu et al. 2019; Li et al. 2020; Tang et al. 2021). Xu et al. (2019), used a three-dimensional deep learning model to segment infected regions in a CT image. The segmented region was categorized into three groups: COVID-19, influenza – a viral pneumonia, and normal using a classification model based on ResNet and location attention structures (Xu et al. 2019).

Researchers have also proposed the application of computer audition systems and artificial intelligence to examine the cough sounds of patients diagnosed with COVID-19 under different conditions such as breathing, dry and wet coughing or sneezing, speaking during colds, eating habits, drowsiness or pain (Schuller et al. 2020). Thereafter, the computer audition system is then used to diagnose and predict the treatment of patients with COVID-19 (Schuller et al. 2020). The respiratory patterns of COVID-19 patients, as well as the breathing patterns of patients with influenza and common colds, have been studied (Wang et al. 2020). The Israeli defense ministry started an investigation on the application of artificial intelligence for the remote diagnosis and monitoring of COVID-19. An application developed by Vocalis Health was used to analyze speech sounds, thereby allowing hospitals to remotely diagnose COVID-19 infected patients without the risk of infection to the healthcare workers.

Blood testing and nucleic acid amplification have been used for the diagnosis of SARS-COV- 2, using machine language classification model for blood screening to extract regular hematological and biochemical parameters, and to provide COVID-19 classification (Wu et al. 2020).

2.3 DRUG DISCOVERY

Another aspect where artificial intelligence proved to be efficient in the management of cold medicine is through the discovery of drugs. The development of potential drugs is an effective way of reducing the mortality rate of COVID-19. Artificial intelligence has been essential in the discovery of COVID-19 vaccines and even in the identification of old vaccine candidates that can be repurposed into new vaccines and therapeutics. The utilization of artificial intelligence in the discovery of drugs and drug repurposing increased rapidly due to the COVID-19 pandemic. Out of

31 studies surveyed by Kaushal et al. (2020), 16 used artificial intelligence for drug repurposing, 10 used artificial intelligence for the discovery and four used artificial intelligence for the development of a vaccine of which only one study developed a stable antibody against the SARS-CoV-2 virus. The study revealed that drug repurposing was the most common application of artificial intelligence for COVID-19 drug discovery (Kaushal et al. 2020). Some examples of repurposed drugs used for the management of COVID-19 are antimalarial drugs such as hydroxychloroquine and antiviral remdesivir and favipiravir (Shrestha et al. 2020). Intelligence technologies sped up the discovery process of drugs through the identification of potential old drug candidates (Monteleone et al. 2021).

Monteleone et al. (2021), defined drug repurposing as a method that identifies new use for a developed drug substance. Drug repurposing is a strategy that is based on two ideas: (1) a single drug can interact with different target sites, thereby laying the foundation for the discovery of new target sites for the drug; (2) target sites of drugs are dependent on a number of different biological processes of pathogenesis (Chavda et al. 2021). The approval of repurposed drug is based on their effectiveness for use rather than the purpose for which the drug was initially established (Mhlanga, 2022).

The development of novel drugs takes a lot of time, from the time of discovery to approval takes approximately 10–15 years and costs a lot of money (2–3 billion US dollars) with a success rate of just 2% (Chavda et al. 2022). Drug development requires five stages which are; drug discovery and development, pre-clinical research, clinical research, FDA review, and FDA post-market safety monitoring and development. This process can, however, be shortened by drug repurposing, which involves only four steps which are; identification of potentially useful compounds, acquisition of the compound, clinical research, FDA postmarket safety and development (Xu et al. 2021).

The approval time of repurposed drug is reduced because the drugs are already in use with familiar side effects. According to Peng et al. (2020), the application of artificial intelligence in the identification of existing drug candidates that can be effective against the virus is a viable way to combat the disease, and it saves a lot of time. Also, in order to reduce the risk of side effects, artificial intelligence can be used to adjust the therapeutic doses of repurposed drugs to suit new diseases as repurposed drugs need to be adapted to the new disease unto which they are being used (Liang et al. 2021).

Different artificial intelligence algorithms that may be useful for drug discovery are alphafold, DeepChen, MOSES, Biopandas, Chemprop, OpenChem, ReLeaSE, neural fingerprint, HTMD, Gnina, ChemGAN challenge, DeepNeuralNet – QSAR, NNScore, SIEVE – score, Hit Dexter (Arora et al. 2020). The commonly used artificial intelligence algorithms are artificial neural network (ANN), convolutional neural network (CNN) and feedforward neural network (FNN) (Monteleone et al. 2021). Both CNN and residual neural network (RNN) have been combined to predict possible antiviral agent drugs that can be remodeled against COVID-19 (Beck et al. 2020). The study identified six antiviral drugs with potential activity against COVID-19.

Several artificial intelligence algorithms have been used to predict useful drug or compound that could be effective against the SARS-CoV-2 virus. Ke et al. (2020), developed an artificial intelligence model to identify possible drug candidates that are effective against COVID-19 using two learning databases. The first databases included components such as compounds reported to be effective against different viruses such as SARS-CoV, SARS-CoV-2, human immunodeficiency virus (HIV), and influenza virus. The second database consisted of 3C-like protease inhibitors. The model identified 80 potential drugs that could be active against the virus, among which eight drugs – bedaquiline, brequinar, coleclxib, clofazimine, conivaptan, gemcitabine, tolcapone, and vizmodegib – were efficient against feline infectious peridonties coronavirus (Ke et al., 2020). Hu et al. (2020), also reported the ability of Almitrine mesylate and Roflumilast used to treat respiratory diseases to be effective against COVID-19 using artificial intelligence.

In another study, Beck et al. (2020), proposed the use of a pre-trained deep learning based drug target interacting (DTI) system called molecule transformer drug target interaction (MT-DTI) which was previously proposed by Boggun et al. (2019). Boggun et al. (2019) collected the amino acid sequences of several components of the viral replication complex, including 3C-like proteinase, RNA-dependent RNA polymerase, helicase, 3′-to-5′ exonuclease, EndoRNase, and 2′-O-ribose methyltransferase, as well as drug targets from NCBI.

Furthermore, a molecular docking and virtual screening tool (Autodeck Vina) (Trot and Olson 2010) was used to predict the possible binding affinity of 3,410 drugs to SARS-CoV-2 3cLpro. The model identified atazanalir, remdesir, efavirenz, ritonavir, and dolutegravir as potential drugs against the SARS-CoV-2 virus (Beck et al. 2020). The MT-DTI model uses simplified molecular-input line-entry system (SMILES) strings and amino acid (AA) sequences to determine the target protein with 3D crystal structure (Beck et al. 2020). Benevolent AI, a startup company that uses artificial intelligence for drug development identified Baricitinid, a drug used for rheumatoid arthritis, as a prospective drug for the treatment of severe COVID-19 symptoms. Following this discovery Eli Lilly, maker of Baricitinid, collaborated with the US national institute of allergy and infectious diseases. The study showed that the clinical trials were effective (Richardson et al. 2020).

Determining the structure of protein is very important in vaccine discovery as the function of protein is defined by its structure and an appropriate protein structure is necessary to ensure essential neutralizing epitopes are present in a way that is recognizable by the B cells (Thomas et al. 2022). The application of artificial intelligence in the development of vaccines includes the prediction of antigen specificity using deep learning algorithms and the identification of small molecule compounds effective against the virus using machine learning (Asada et al., 2021). The application of artificial intelligence and machine learning in the development of vaccines is based on the prediction of possible epitopes using different artificial intelligence algorithms such as ANN, gradient boosting decision trees and deep neural networks.

In a study by Fast et al. (2020), two previously developed ANN models called MARIA and NetMHCPan4 were used to identify both T-cell and B-cell epitopes of the SARS-CoV-2 virus. The algorithm identified 405 T-cell epitopes with a strong presentation score for both major histocompatibility complexes (i.e. MHC 1 and MHC II) and also two potential neutralizing epitopes on the S-protein. The epitopes predicted in this study maybe be useful in the development of COVID-19 vaccine. Ong et al. (2020) identified five non-structural proteins (NSPs) among which non-structural protein 3 (nsp3) and non-structural protein 8 (nsp8) were predicted the best vaccine target site using a machine learning framework.

2.4 EPIDEMIOLOGY AND SOCIAL CONTROL

Artificial intelligence can be applied in the monitoring and management of the spread of COVID-19. In this context artificial intelligence can be used to predict the spread of the disease, identify high-risk patients and identify possible control measures (Senthilraja, 2021). Senthilraja (2021), acknowledged the ability of artificial intelligence to track and forecast the nature of the SARS-CoV-2 from big data available from various platforms such as social media, public health surveillance and predict the rates at which the virus spreads. The application of artificial intelligence in epidemic control and management includes video tracking, individual temperature detection, contact tracing, and intelligent robots (Chen and See, 2020). Artificial intelligence-based algorithms are important in the development of epidemiological spread model that could be useful in the fight against the pandemic (Musulin et al. 2021).

According to Mhlanga (2022), the COVID-19 virus was first discovered by a forecasting application called BlueDot. BlueDot, a Canadian-based startup company, predicted the outbreak of the infection by issuing a warning to its clients on December 30, 2019. This alert from BlueDot came ten days before the World Health Organization (WHO) warning on January 9, 2020. Twenty destination cities where passengers from Wuhan would arrive at the beginning of the outbreak were predicted by Bogoch et al. (2022). The researchers warned that the listed cities would be at the center of the global spread of the disease. However, another artificial intelligence model, HealthMap, issued an alert about the outbreak on December 30, 2018, seven days earlier than the warning from the United States Centers for Disease Control and Prevention (USCDC). Li et al. (2020), proposed a model that analyzes the data on the Hubei epidemic using big data. The authors were able to determine the evolution trend of the virus in countries such as south Korea, Iran, and Italy through available data. Furthermore, the study came up with controls that were important in the pandemic.

One of the most common method of determining the spread of COVID-19 is by using the long- and short-term memory (LSTM) network due to its ability to perform better in longer sequences compared to normal RNNs. (Peng et al., 2020). Tomar and Gupta (2020) used the LSTM network to forecast the number of recovered cases, daily positive cases, and death cases, as well as develop applicable preventive measure. Chimmula et al. (2020) have also used the LSTM network in the management and spread prediction of COVID-19. Other models that have, however, been used to predict to predict the spread of the virus include the logistic growth model (Chen and See, 2020), TP-SMN-AR model (Maleki et al. 2020), machine learning model (Tiwari et al. 2020), four time series model (Al-Rousan and Al-Najjar, 2020), Seir model and artificial intelligence approach (Yang et al. 2020), group method of data handling (GMDH) algorithms (Pirouz et al. 2020).

Yahya et al. (2020), built a framework to determine the spread of COVID-19 in Iraq using the infection data supplied by the Iraqi ministry of health. The model used three ANN functions (radial-based neurons, NARX neurons, FC-M neurons). The model achieved a performance accuracy of 81.6%. Tamang et al. (2020), developed a forecasting model based on ANN to predict future cases of COVID-19 outbreak in India, USA, France, and UK. The study predicted that the COVID-19 cases in India, USA, France, and the UK will rise from about 50,000 to 160,000, 1,200,000 to 1,700,00, 140,000 to 150,000 and 240,000 to 250,000 in two to ten months based on the trend shown in China and South Korea. Furthermore, the study predicted the death toll of these countries to be about 1,600 to 4,000 for India, 135,000 to 100,000 for USA, 40,000 to 50, 000 for France and 35,000 to 47,000 for UK in two to ten months.

Mollanlo et al. (2020), developed an ANN model to predict the incidence rate of COVID-19 across the continental region of USA. Sujath et al. (2020), proposed an artificial intelligence model to predict the spread of COVID-19 in India using data collected from Kaggle. Kaggle is an online platform where people can find and work with data, participate in data science competitions, and share their projects and insights. It provides datasets for analysis, coding challenges, and a community for learning and collaboration in data science and machine learning.

Kafieh et al. (2020), designed an agent-based artificial intelligence simulation platform (Enerpol), to forecast the spread of COVID-19 and determine a containment method for the outbreak. Osman et al. (2021), used a topological autoencoder to generate a similarity map to predict the spread of COVID-19. Rica and Ruz, (2020), used a differential evolution with the SEIR-SD model to determine the spread of the COVID-19 virus. Pabat and Chakraborty (2020), used a machine learning model (SVR) to predict the number of confirmed, deceased and recovered COVID-19 cases using data collected from Kaggle.

After the diagnosis of an infected individual, the next important step to prevent the further spread of the disease is contact tracing (Mhlanga, 2022). Contact tracing is a strategy that identifies individuals who had been in close contact with the infected carrier as they maybe probably infected (Ahmed et al. 2020). This is an important public health tool in order to break the chain of transmission and control the spread of the SARS-CoV-2 virus (Mhlanga, 2022). This is usually achieved through the interview of infected individual by public health authorities (Ahmed et al. 2020).

Several artificial intelligence technologies were developed and used by many countries in the course of the pandemic. Bluetooth and geofencing were integrated into different contact tracing applications and deployed in many countries. Before the outbreak of COVID-19, a geofencing application was used as a marketing tool by alerting people about nearby stores and products (Mhlanga, 2022). Hui (2020) reported that that geofencing was implemented in China to monitor quarantined COVID-19 patients. It was also used to mark infected areas and alert health authorities Culham (2021).

Table 2.1 shows the application of artificial intelligence in the management COVID-19 by different countries.

TABLE 2.1
The Application of Artificial Intelligence in the Management COVID-19 by Different Countries

Countries	AI Application	Function
Afghanistan	ASAN	Information, tracking
Angola	COVID-19 AO	Self-diagnosis and quarantine enforcement
Argentina	CUIDAR	Contact tracing
Armenia	ArMed eHealth	Contact tracing
Armenia	Covid-19 Armenia	COVID-19 information
Australia	Coronavirus Australia	COVID-19 information, isolation registration
Australia	COVIDSafe	Contact tracing
Austria	Stopp Corona	Contact tracing
Azerbaijan	e-Tabib	Information
Bahrain	BeAware Bahrain	Contact tracing
Bangladesh	Corona Tracer BD	Contact tracing
Bangladesh	Surokkha	Vaccination, Contact Tracing
Belgium	Coronalert	Contact tracing
Bhutan	Druk Trace	Contact tracing
Bolivia	Unidos Contra el Covid	Contact tracing
Brazil	Coronavírus – SUS	Contact tracing
Brazil	Tô de Olho	Contact tracing
Brazil	VirusMapBR (formerly The Spread Project)	Contact tracing, medical reporting
Brunei	BruHealth	Contact tracing
Bulgaria	ViruSafe	Contact tracing
Cambodia	Khmer Vacc	Vaccination, medical response
Canada	ABTraceTogether	Contact tracing
Canada	COVID Alert	Contact tracing
Canada	TeamSense	Covid screening for employees.
Chile	CoronApp - Chile	Information
China	"Health Code"	Contact tracing
Colombia	CoronApp – Colombia	Information
Cyprus	CovTracer	Contact tracing and warning
Czech Republic	eRouška	Contact tracing
Denmark	smittelstop	Contact tracing
Ecuador	ASI	Contact tracing
Ecuador	CovidEC	Information
Egypt	Egypt Health Passport app	Information
Egypt	Sehet Misr	Contact tracing
Fiji	careFIJI	Contact tracing
Finland	Koronavilkku	Contact tracing
France	Check Your Mask	Checking if masks are worn correctly

(Continued)

TABLE 2.1 (CONTINUED)

Countries	AI Application	Function
France	ROBERT (robust and privacy-preserving proximity tracing protocol)	Contact tracing
France	TousAntiCovid	Contact tracing
Georgia	Stop Covid	Contact tracing
Germany	Corona-Warn-App	Contact tracing
Germany	Ito	Contact tracing
Germany	OHIOH Research	Contact tracing, scientific research
Ghana	GH COVID-19 Tracker App	Contact tracing
Greece	DOCANDU Covid Checker	Self-diagnostic, online doctor
Hong Kong	LeaveHomeSafe	Contact tracking
Hong Kong	StayHomeSafe	Quarantine enforcement
Hungary	VírusRadar	Contact tracing
India	Aarogya Setu	Contact tracing
India	Corona Kavach	Information
India	COVA Punjab	Contact tracing
India	COVID-19 Quarantine Monitor	Contact tracing
Indonesia	PeduliLindungi	Contact tracing
Iran	AC19	Contact tracing
Ireland	COVID tracker Ireland	Contact tracing, health information EU Digital Vaccination Certificate
Israel	Hamagen	Contact tracing
Italy	Covid Community Alert	Contact tracing
Italy	SM-COVID-19	Contact Tracing
Kazakhstan	Ashyq	Contact tracing
Kuwait	Shlonik	Contact tracing
Kyrgyzstan	STOP COVID-19	contact tracing
Laos	LaoKYC	Contact tracing
Latvia	Apturi Covid	Exposure Detection
Lebanon	MOPH	Information
Luxembourg	CovidCheck.lu	Contact tracing
Malaysia	Gerak Malaysia	Contact tracing
Malaysia	MyTrace	Contact tracing
Maldives	TraceEkee	Contact tracing
Malta	COVID Alert Malta	Contact tracing
Monaco	Safe Pass	Information
Mongolia	Ersdel	Exposure Detection
Morocco	Wiqaytna	Contact tracing
Myanmar	Saw Saw Shar	Information
Netherlands	CoronaMelder	Contact tracing
Netherlands	PrivateTracer	Contact tracing
New Zealand	NZ COVID Tracer	Contact tracing
North Macedonia	StopKorona	Contact tracing
Norway	Smittestopp	Contact tracing
Pakistan	COVID-19 Gov PK	Contact tracing

2.5 CONCLUSION

Artificial intelligence has been applied in every aspect of the fight against COVID-19. It's application is important in the development of easier, faster, accurate readily available diagnostic methods. It is also applied in drug discovery in predicting the structure of proteins, and identifying potential drug candidates that can be repurposed against the virus. In epidemiology, artificial intelligence can be applied in forecasting the spread of the virus, developing measures that may be useful in preventing the spread of the virus, and also for contact tracing of infected individuals.

REFERENCES

Abd-Alrazaq, A., Alajlani, M., Alhuwail, D., Schneider, J., Al-Kuwari, S., Shah, Z., Hamdi, M., & Househ, M. (2020). Artificial intelligence in the fight against COVID-19: Scoping review. *Journal of Medical Internet Research, 22*(12), e20756.

Abraham, B., & Nair, M. S. (2020). Computer-aided detection of COVID-19 from X-ray images using multi-CNN and Bayesnet classifier. *Biocybernetics and Biomedical Engineering, 40*(4), 1436–1445.

Adetunji, Charles Oluwaseun, Egbuna, Chukwuebuka, Oladosun, Tolulope Olawumi, Akram, Muhammad, Michael, Olugbenga, Olisaka, Frances Ngozi, Ozolua, Phebean, Adetunji, Juliana Bunmi, Enoyoze, Goddidit Esiro, & Olaniyan, Olugbemi (2021). Efficacy of phytochemicals of medicinal plants for the treatment of human echinococcosis. In *Neglected tropical diseases and phytochemicals in drug discovery*. DOI: 10.1002/9781119617143.ch8

Adetunji, Charles Oluwaseun, Mitembo, William Peter, Egbuna, Chukwuebuka, Narasimha Rao, G.M. (2020). In silico modeling as a tool to predict and characterize plant toxicity. In Andrew G. Mtewa, Chukwuebuka Egbuna, & G.M. Narasimha Rao (Eds.), *Poisonous plants and phytochemicals in drug discovery*. Wiley Online Library DOI: 10.1002/9781119650034.ch14

Adetunji, Charles Oluwaseun, Nwankwo, Wilson, Olayinka, Akinola Samson, Olugbemi, Olaniyan Tope, Akram, Muhammad, Laila, Umme, Olugbenga, Michael Samuel, Oshinjo, Ayomide Michael, Adetunji, Juliana Bunmi, Okotie, Gloria E., & Esiobu, Nwadiuto (Diuto) (2022h). Machine learning and behaviour modification for COVID-19. In *Medical biotechnology, biopharmaceutics, forensic science and bioinformatics* (1st ed.). First Published 2022. Imprint CRC Press. Pages 17. eBook ISBN 9781003178903. DOI: 10.1201/9781003178903-17

Adetunji, Charles Oluwaseun, Nwankwo, Wilson, Olayinka, Akinola Samson, Olugbemi, Olaniyan Tope, Akram, Muhammad, Laila, Umme, Samuel, Michael Olugbenga, Oshinjo, Ayomide Michael, Adetunji, Juliana Bunmi, Okotie, Gloria E., & Esiobu, Nwadiuto (Diuto) (2022f). Computational intelligence techniques for combating COVID-19. In *Medical biotechnology, biopharmaceutics, forensic science and bioinformatics* (1st ed.). First Published 2022. Imprint CRC Press. Pages 12. eBook ISBN 9781003178903. DOI: 10.1201/9781003178903-16

Adetunji, Charles Oluwaseun, Olaniyan, Olugbemi Tope, Adeyomoye, Olorunsola, Dare, Ayobami, Adeniyi, Mayowa J, Alex, Enoch, Rebezov, Maksim, Garipova, Larisa, & Shariati, Mohammad Ali. (2022a). eHealth, mHealth, and telemedicine for COVID-19 pandemic. In S.K. Pani, S. Dash, W.P. dos Santos, S.A. Chan Bukhari, & F. Flammini (Eds.), *Assessing COVID-19 and other pandemics and epidemics using computational modelling and data analysis*. Springer. DOI: 10.1007/978-3-030-79753-9_10

Adetunji, Charles Oluwaseun, Olaniyan, Olugbemi Tope, Adeyomoye, Olorunsola, Dare, Ayobami, Adeniyi, Mayowa J., Alex, Enoch, Rebezov, Maksim, Isabekova, Olga, & Shariati, Mohammad Ali (2022c). Smart sensing for COVID-19 pandemic. In S.K. Pani, S. Dash, W.P. dos Santos, S.A. Chan Bukhari, & F. Flammini (Eds.), *Assessing COVID-19 and other pandemics and epidemics using computational modelling and data analysis*. Springer. DOI: 10.1007/978-3-030-79753-9_9

Adetunji, Charles Oluwaseun, Olaniyan, Olugbemi Tope, Adeyomoye, Olorunsola, Dare, Ayobami, Adeniyi, Mayowa J., Alex, Enoch, Rebezov, Maksim, Koriagina, Natalia, & Shariati, Mohammad Ali (2022e). Diverse techniques applied for effective diagnosis of COVID-19. In S.K. Pani, S. Dash, W.P. dos Santos, S.A. Chan Bukhari, & F. Flammini (Eds.), *Assessing COVID-19 and other pandemics and epidemics using computational modelling and data analysis*. Springer. DOI: 10.1007/978-3-030-79753-9_3

Adetunji, Charles Oluwaseun, Olaniyan, Olugbemi Tope, Olorunsola Adeyomoye, Ayobami Dare, Mayowa J. Adeniyi, Enoch Alex, Maksim Rebezov, Ekaterina Petukhova, & Shariati, Mohammad Ali (2022b). Machine learning approaches for COVID-19 pandemic. In S.K. Pani, S. Dash, W.P. dos Santos, S.A. Chan Bukhari, & F. Flammini (Eds.), *Assessing COVID-19 and other pandemics and epidemics using computational modelling and data analysis*. Springer. DOI: 10.1007/978-3-030-79753-9_8

Adetunji, Charles Oluwaseun, Olaniyan, Olugbemi Tope, Adeyomoye, Olorunsola, Dare, Ayobami, Adeniyi, Mayowa J., Alex, Enoch, Rebezov, Maksim, Petukhova, Ekaterina, & Shariati, Mohammad Ali (2022d). Internet of Health Things (IoHT) for COVID-19. In S.K. Pani, S. Dash, W.P. dos Santos, S.A. Chan Bukhari, & F. Flammini (Eds.), *Assessing COVID-19 and other pandemics and epidemics using computational modelling and data analysis*. Springer. DOI: 10.1007/978-3-030-79753-9_5

Adetunji, Charles Oluwaseun, Olugbemi, Olaniyan Tope, Akram, Muhammad, Laila, Umme, Samuel, Michael Olugbenga, Oshinjo, Ayomide Michael, Adetunji, Juliana Bunmi, Okotie, Gloria E., Esiobu, Nwadiuto (Diuto), Oyedara, Omotayo Opemipo, & Adeyemi, Folasade Muibat (2022g). Application of computational and bioinformatics techniques in drug repurposing for effective development of potential drug candidate for the management of COVID-19. In *Medical biotechnology, biopharmaceutics, forensic science and bioinformatics* (1st ed.). First Published 2022, Imprint CRC Press. Pages 14. eBook ISBN 9781003178903. DOI: 10.1201/9781003178903-15

Adetunji, Charles Oluwaseun, & Oyeyemi, Oyetunde T. (2022). Antiprotozoal activity of some medicinal plants against entamoeba histolytica, the causative agent of amoebiasis. In *Medical biotechnology, biopharmaceutics, forensic science and bioinformatics* (1st ed.). First Published 2022. Imprint CRC Press. Pages 12. eBook ISBN 9781003178903. https://www.taylorfrancis.com/chapters/edit/10.1201/9781003178903-20/antiprotozoal-activity-medicinal-plants-entamoeba-histolytica-causative-agent-amoebiasis-charles-oluwaseun-adetunji-oyetunde-oyeyemi

Adly, A. S., Adly, A. S., & Adly, M. S. (2020). Approaches based on artificial intelligence and the internet of intelligent things to prevent the spread of COVID-19: Scoping review. *Journal of Medical Internet Research*, *22*(8), e19104.

Ahmed, N., Michelin, R. A., Xue, W., Ruj, S., Malaney, R., Kanhere, S. S., Seneviratne, A., Hu, W., Janicke, H., & Jha, S. K. (2020). A survey of COVID-19 contact tracing apps. *IEEE Access*, *8*, 134577–134601.

Ahuja, A. S., Reddy, V. P., & Marques, O. (2020). Artificial intelligence and COVID-19: A multidisciplinary approach. *Integrative Medicine Research*, *9*(3), 100434. https://doi.org/10.1016/j.imr.2020.100434

Akram, Muhammad, Adetunji, Charles Oluwaseun, Egbuna, Chukwuebuka, Jabeen, Shaista, Olaniyan, Olugbemi, Ezeofor, Nebechi Jane, Anani, Osikemekha Anthony, Laila, Umme, Găman, Mihnea-Alexandru, Patrick-Iwuanyanwu, Kingsley, Ifemeje, Jonathan Chinenye, Chikwendu, Chukwudi Jude, Michael, Olisah Chinedu, & Rudrapal, Mithun (2021). Dengue fever. In *Neglected tropical diseases and phytochemicals in drug discovery*. https://doi.org/10.1002/9781119617143.ch17

Akram, Muhammad, Mohiuddin, Ejaz, Adetunji, Charles Oluwaseun, Oladosun, Tolulope Olawumi, Ozolua, Phebean, Olisaka, Frances Ngozi, Egbuna, Chukwuebuka, Michael, Olugbenga, Adetunji, Juliana Bunmi, Hameed, Leena, Awuchi, Chinaza Godswill, Patrick-Iwuanyanwu, Kingsley, & Olaniyan, Olugbemi (2021). Chapter 7: Prospects of phytochemicals for the treatment of helminthiasis. In *Neglected tropical diseases and phytochemicals in drug discovery* (1st ed.). Wiley. https://doi.org/10.1002/9781119617143.ch7

Albahri, O. S., Zaidan, A. A., Albahri, A. S., Zaidan, B. B., Abdulkareem, K. H., Al-Qaysi, Z. T., Alamoodi, A. H., Aleesa, A. M., Chyad, M. A., & Alesa, R. M. (2020). Systematic review of artificial intelligence techniques in the detection and classification of COVID-19 medical images in terms of evaluation and benchmarking: Taxonomy analysis, challenges, future solutions and methodological aspects. *Journal of Infection and Public Health, 13*(10), 1381–1396.

Alimadadi, A., Aryal, S., Manandhar, I., Munroe, P. B., Joe, B., & Cheng, X. (2020). Artificial intelligence and machine learning to fight COVID-19. In *Physiological genomics, 52*(4), 200–202. https://doi.org/10.1152/physiolgenomics.00029.2020

Al-Qaness, M. A., Saba, A. I., Elsheikh, A. H., Abd Elaziz, M., Ibrahim, R. A., Lu, S., Hemedan, A. A., Shanmugan, S., & Ewees, A. A. (2021). Efficient artificial intelligence forecasting models for COVID-19 outbreak in Russia and Brazil. *Process Safety and Environmental Protection, 149*, 399–409.

Al-Rousan, N., & Al-Najjar, H. (2020). The impact of Chinese government plans on coronavirus CoVID-19 spreading and its association with weather variables in 30 Chinese provinces: Investigation. *European Review for Medical and Pharmacological Sciences, 24*(8), 4565–4571.

Arora, K., & Bist, A. S. (2020). Artificial intelligence based drug discovery techniques for covid-19 detection. *Aptisi Transactions on Technopreneurship (ATT), 2*(2), 120–126.

Arora, N., Banerjee, A. K., & Narasu, M. L. (2020). The role of artificial intelligence in tackling COVID-19. In *Future virology*, 15(11), 717–724. https://doi.org/10.2217/fvl-2020-0130

Asada, K., Komatsu, M., Shimoyama, R., Takasawa, K., Shinkai, N., Sakai, A., Bolatkan, A., Yamada, M., Takahashi, S., & Machino, H. (2021). Application of artificial intelligence in COVID-19 diagnosis and therapeutics. *Journal of Personalized Medicine, 11*(9), 886.

Ballı, S. (2021). Data analysis of Covid-19 pandemic and short-term cumulative case forecasting using machine learning time series methods. *Chaos, Solitons & Fractals, 142*, 110512.

Bansal, A., Padappayil, R. P., Garg, C., Singal, A., Gupta, M., & Klein, A. (2020). Utility of artificial intelligence amidst the COVID 19 pandemic: A review. *Journal of Medical Systems, 44*(9), 1–6.

Beck, M. J., & Hensher, D. A. (2020). Insights into the impact of COVID-19 on household travel and activities in Australia–The early days of easing restrictions. *Transport Policy, 99*, 95–119.

Belfiore, M. P., Urraro, F., Grassi, R., Giacobbe, G., Patelli, G., Cappabianca, S., & Reginelli, A. (2020). Artificial intelligence to codify lung CT in Covid-19 patients. *La Radiologia Medica, 125*(5), 500–504.

Boggan, R. M., Lim, A., Taylor, R. W., McFarland, R., & Pickett, S. J. (2019). Resolving complexity in mitochondrial disease: Towards precision medicine. *Molecular Genetics and Metabolism, 128*(1–2), 19–29. https://doi.org/10.1016/j.ymgme.2019.09.003

Bogoch, I. I., & Halani, S. (2022). COVID-19 vaccines: A geographic, social and policy view of vaccination efforts in Ontario, Canada. *Cambridge Journal of Regions, Economy and Society, 15*(3), 757–770.

Borkowski, A. A., Viswanadhan, N. A., Thomas, L. B., Guzman, R. D., Deland, L. A., & Mastorides, S. M. (2020). Using artificial intelligence for COVID-19 chest X-ray diagnosis. *Federal Practitioner, 37*(9), 398.

Bouchareb, Y., Khaniabadi, P. M., Al Kindi, F., Al Dhuhli, H., Shiri, I., Zaidi, H., & Rahmim, A. (2021). Artificial intelligence-driven assessment of radiological images for COVID-19. *Computers in Biology and Medicine, 136*, 104665.

Bragazzi, N. L., Dai, H., Damiani, G., Behzadifar, M., Martini, M., & Wu, J. (2020). How big data and artificial intelligence can help better manage the COVID-19 pandemic. *International Journal of Environmental Research and Public Health*, *17*(9), 3176.

Browning, A., Moss, M. E., & Berkman, E. (2021). *Leveraging evidence-based messaging to prevent the spread of COVID-19*. https://osf.io/preprints/psyarxiv/thm2w/

Bullock, J., Luccioni, A., Pham, K. H., Lam, C. S. N., & Luengo-Oroz, M. (2020). Mapping the landscape of artificial intelligence applications against COVID-19. *Journal of Artificial Intelligence Research*, *69*, 807–845.

Chang, A. C. (2020). Artificial intelligence and COVID-19: Present state and future vision. *Intelligence-Based Medicine*, *3*, 100012.

Chavda, V. P. (2023). Artificial Intelligence and Machine Learning-Based Formulation and Process Development for Drug Products. In V. Chavda, K. Anand, & V. Apostolopoulos (Eds.), *Bioinformatics Tools for Pharmaceutical Drug Product Development* (1st ed., pp. 183–195). Wiley. https://doi.org/10.1002/9781119865728.ch9

Chavda, V. P., Gajjar, N., Shah, N., & Dave, D. J. (2021). Darunavir ethanolate: Repurposing an anti-HIV drug in COVID-19 treatment. *European Journal of Medicinal Chemistry Reports*, *3*, 100013.

Chen, J., Li, K., Zhang, Z., Li, K., & Yu, P. S. (2021). A survey on applications of artificial intelligence in fighting against COVID-19. *ACM Computing Surveys (CSUR)*, *54*(8), 1–32.

Chen, J., & See, K. C. (2020). Artificial intelligence for COVID-19: Rapid review. *Journal of Medical Internet Research*, *22*(10), e21476.

Chimmula, V. K. R., & Zhang, L. (2020). Time series forecasting of COVID-19 transmission in Canada using LSTM networks. *Chaos, Solitons & Fractals*, *135*, 109864.

Culham, D. (n.d.). *Utilizing Additive Manufacturing to Digitize Warehouses*. Retrieved September 15, 2024, from https://cs-strategies.com/covid-19-utilizing-additive-manufacturing-to-digitize-warehouses/

Dananjayan, S., & Raj, G. M. (2020). Artificial Intelligence during a pandemic: The COVID-19 example. *The International Journal of Health Planning and Management*, *35*(5), 1260–1262. https://doi.org/10.1002/hpm.2987

Fast, E., Altman, R. B., & Chen, B. (2020). Potential T-cell and B-cell epitopes of 2019-nCoV. *BioRxiv*, https://doi.org/10.1101/2020.02.19.955484

Gunasekeran, D. V., Tseng, R. M. W. W., Tham, Y.-C., & Wong, T. Y. (2021). Applications of digital health for public health responses to COVID-19: A systematic scoping review of artificial intelligence, telehealth and related technologies. *NPJ Digital Medicine*, *4*(1), 1–6.

Harmon, S. A., Sanford, T. H., Xu, S., Turkbey, E. B., Roth, H., Xu, Z., Yang, D., Myronenko, A., Anderson, V., & Amalou, A. (2020). Artificial intelligence for the detection of COVID-19 pneumonia on chest CT using multinational datasets. *Nature Communications*, *11*(1), 1–7.

Health, T. L. D. (2021). Artificial intelligence for COVID-19: Saviour or saboteur? *The Lancet. Digital Health*, *3*(1), e1.

Ho, D. (2020). Addressing COVID-19 drug development with artificial intelligence. *Advanced Intelligent Systems*, *2*(5), 2000070.

Hu, Z., Ge, Q., Li, S., Jin, L., & Xiong, M. (2020). Artificial intelligence forecasting of Covid-19 in China. *ArXiv Preprint ArXiv:2002.07112*.

Huang, S., Yang, J., Fong, S., & Zhao, Q. (2021). Artificial intelligence in the diagnosis of COVID-19: Challenges and perspectives. *International Journal of Biological Sciences*, *17*(6), 1581.

Hui, D. S., Azhar, E. I., Madani, T. A., Ntoumi, F., Kock, R., Dar, O., Ippolito, G., Mchugh, T. D., Memish, Z. A., & Drosten, C. (2020). The continuing 2019-nCoV epidemic threat of novel coronaviruses to global health—The latest 2019 novel coronavirus outbreak in Wuhan, China. *International Journal of Infectious Diseases*, *91*, 264–266.

Hung, M., Lauren, E., Hon, E. S., Birmingham, W. C., Xu, J., Su, S., Hon, S. D., Park, J., Dang, P., & Lipsky, M. S. (2020). Social network analysis of COVID-19 sentiments: Application of artificial intelligence. *Journal of Medical Internet Research*, *22*(8), e22590.

Hussain, A., Tahir, A., Hussain, Z., Sheikh, Z., Gogate, M., Dashtipour, K., Ali, A., & Sheikh, A. (2021). Artificial intelligence–enabled analysis of public attitudes on Facebook and Twitter toward Covid-19 vaccines in the united kingdom and the united states: Observational study. *Journal of Medical Internet Research*, *23*(4), e26627.

Islam, M. M., Poly, T. N., Alsinglawi, B., Lin, L.-F., Chien, S.-C., Liu, J.-C., & Jian, W.-S. (2021). Application of artificial intelligence in COVID-19 pandemic: Bibliometric analysis. *Healthcare*, *9*(4), 441. https://www.mdpi.com/2227-9032/9/4/441

Jamshidi, M., Lalbakhsh, A., Talla, J., Peroutka, Z., Hadjilooei, F., Lalbakhsh, P., Jamshidi, M., La Spada, L., Mirmozafari, M., & Dehghani, M. (2020). Artificial intelligence and COVID-19: Deep learning approaches for diagnosis and treatment. *IEEE Access*, *8*, 109581–109595.

Jin, C., Chen, W., Cao, Y., Xu, Z., Tan, Z., Zhang, X., Deng, L., Zheng, C., Zhou, J., & Shi, H. (2020). Development and evaluation of an artificial intelligence system for COVID-19 diagnosis. *Nature Communications*, *11*(1), 1–14.

Kafieh, R., Arian, R., Saeedizadeh, N., Amini, Z., Serej, N. D., Minaee, S., Yadav, S. K., Vaezi, A., Rezaei, N., & Haghjooy Javanmard, S. (2021). COVID-19 in Iran: Forecasting Pandemic Using Deep Learning. *Computational and Mathematical Methods in Medicine*, *2021*, 1–16. https://doi.org/10.1155/2021/6927985

Kaushal, J., & Mahajan, P. (2021). Asia's largest urban slum-Dharavi: A global model for management of COVID-19. *Cities*, *111*, 103097.

Ke, Y.-Y., Peng, T.-T., Yeh, T.-K., Huang, W.-Z., Chang, S.-E., Wu, S.-H., Hung, H.-C., Hsu, T.-A., Lee, S.-J., Song, J.-S., Lin, W.-H., Chiang, T.-J., Lin, J.-H., Sytwu, H.-K., & Chen, C.-T. (2020). Artificial intelligence approach fighting COVID-19 with repurposing drugs. *Biomedical Journal*, *43*(4), 355–362. https://doi.org/10.1016/j.bj.2020.05.001

Keshavarzi Arshadi, A., Webb, J., Salem, M., Cruz, E., Calad-Thomson, S., Ghadirian, N., Collins, J., Diez-Cecilia, E., Kelly, B., & Goodarzi, H. (2020). Artificial intelligence for COVID-19 drug discovery and vaccine development. *Frontiers in Artificial Intelligence*, *3*, 65.

Khan, M., Mehran, M. T., Haq, Z. U., Ullah, Z., Naqvi, S. R., Ihsan, M., & Abbass, H. (2021). Applications of artificial intelligence in COVID-19 pandemic: A comprehensive review. *Expert Systems with Applications*, *185*, 115695.

Laghi, A. (2020). Cautions about radiologic diagnosis of COVID-19 infection driven by artificial intelligence. *The Lancet Digital Health*, *2*(5), e225.

Laguarta, J., Hueto, F., & Subirana, B. (2020). COVID-19 artificial intelligence diagnosis using only cough recordings. *IEEE Open Journal of Engineering in Medicine and Biology*, *1*, 275–281.

Lalmuanawma, S., Hussain, J., & Chhakchhuak, L. (2020). Applications of machine learning and artificial intelligence for Covid-19 (SARS-CoV-2) pandemic: A review. *Chaos, Solitons & Fractals*, *139*, 110059.

Li, L., Qin, L., Xu, Z., Yin, Y., Wang, X., Kong, B., Bai, J., Lu, Y., Fang, Z., & Song, Q. (2020). Artificial intelligence distinguishes COVID-19 from community acquired pneumonia on chest CT. *Radiology*, 200905. https://doi.org/10.1148/radiol.2020200905

Liang, S., Liu, C., Rotaru, K., Li, K., Wei, X., Yuan, S., Yang, Q., Ren, L., & Liu, X. (2022). The relations between emotion regulation, depression and anxiety among medical staff during the late stage of COVID-19 pandemic: A network analysis. *Psychiatry Research*, *317*, 114863.

Liang, W., Liang, H., Ou, L., Chen, B., Chen, A., Li, C., Li, Y., Guan, W., Sang, L., & Lu, J. (2020). Development and validation of a clinical risk score to predict the occurrence of critical illness in hospitalized patients with COVID-19. *JAMA Internal Medicine*, *180*(8), 1081–1089.

Lu, Y. (2019). Artificial intelligence: A survey on evolution, models, applications and future trends. *Journal of Management Analytics*, *6*(1), 1–29. https://doi.org/10.1080/23270012.2019.1570365

Maleki, M., Mahmoudi, M. R., Wraith, D., & Pho, K.-H. (2020). Time series modelling to forecast the confirmed and recovered cases of COVID-19. *Travel Medicine and Infectious Disease*, *37*, 101742.

Malik, Y. S., Sircar, S., Bhat, S., Ansari, M. I., Pande, T., Kumar, P., Mathapati, B., Balasubramanian, G., Kaushik, R., & Natesan, S. (2021). How artificial intelligence may help the Covid-19 pandemic: Pitfalls and lessons for the future. *Reviews in Medical Virology*, *31*(5), 1–11.

McCall, B. (2020). COVID-19 and artificial intelligence: Protecting health-care workers and curbing the spread. *The Lancet Digital Health*, *2*(4), e166–e167.

Mei, X., Lee, H.-C., Diao, K., Huang, M., Lin, B., Liu, C., Xie, Z., Ma, Y., Robson, P. M., & Chung, M. (2020). Artificial intelligence–enabled rapid diagnosis of patients with COVID-19. *Nature Medicine*, *26*(8), 1224–1228.

Mhlanga, D. (n.d.). SSIRC 2023-059 Sustaining the COVID-19 induced digital transformation in education mechanisms for success in the fourth industrial revolution. *ID Paper*, p. 325.

Mohamadou, Y., Halidou, A., & Kapen, P. T. (2020). A review of mathematical modeling, artificial intelligence and datasets used in the study, prediction and management of COVID-19. *Applied Intelligence*, *50*(11), 3913–3925.

Mohanty, S., Rashid, M. H. A., Mridul, M., Mohanty, C., & Swayamsiddha, S. (2020). Application of artificial intelligence in COVID-19 drug repurposing. *Diabetes & Metabolic Syndrome: Clinical Research & Reviews*, *14*(5), 1027–1031.

Mollalo, A., Rivera, K. M., & Vahedi, B. (2020). Artificial Neural Network Modeling of Novel Coronavirus (COVID-19) Incidence Rates across the Continental United States. *International Journal of Environmental Research and Public Health*, *17*(12), 4204. https://doi.org/10.3390/ijerph17124204

Monteleone, A. M., Marciello, F., Cascino, G., Abbate-Daga, G., Anselmetti, S., Baiano, M., Balestrieri, M., Barone, E., Bertelli, S., & Carpiniello, B. (2021). The impact of COVID-19 lockdown and of the following "re-opening" period on specific and general psychopathology in people with Eating Disorders: The emergent role of internalizing symptoms. *Journal of Affective Disorders*, *285*, 77–83.

Murphy, K., Smits, H., Knoops, A. J., Korst, M. B., Samson, T., Scholten, E. T., Schalekamp, S., Schaefer-Prokop, C. M., Philipsen, R. H., & Meijers, A. (2020). COVID-19 on chest radiographs: A multireader evaluation of an artificial intelligence system. *Radiology*, *296*(3), E166–E172.

Musulin, J., Baressi Šegota, S., Štifanić, D., Lorencin, I., Andjelić, N., Šušteršič, T., Blagojević, A., Filipović, N., Ćabov, T., & Markova-Car, E. (2021). Application of artificial intelligence-based regression methods in the problem of COVID-19 spread prediction: A systematic review. *International Journal of Environmental Research and Public Health*, *18*(8), 4287.

Naudé, W. (2020a). *Artificial Intelligence against COVID-19: An early review*. (Working Paper 13110). IZA Discussion Papers. www.econstor.eu/handle/10419/216422

Naudé, W. (2020b). Artificial intelligence vs COVID-19: Limitations, constraints and pitfalls. *AI & Society*, *35*(3), 761–765.

Nawaz, M. S., Fournier-Viger, P., Shojaee, A., & Fujita, H. (2021). Using artificial intelligence techniques for COVID-19 genome analysis. *Applied Intelligence*, *51*(5), 3086–3103.

Obeid, J. S., Davis, M., Turner, M., Meystre, S. M., Heider, P. M., O'Bryan, E. C., & Lenert, L. A. (2020). An artificial intelligence approach to COVID-19 infection risk assessment in virtual visits: A case report. *Journal of the American Medical Informatics Association*, *27*(8), 1321–1325.

Olaniyan, Olugbemi T., Adetunji, Charles O., Adeniyi, Mayowa J., & Hefft, Daniel Ingo (2022a). Machine learning techniques for high-performance Computing for IoT applications in healthcare. In *Deep learning, machine learning and IoT in biomedical and health informatics* (1st ed.). First Published 2022. Imprint CRC Press. Pages 13. eBook ISBN 9780367548445. DOI: 10.1201/9780367548445-20

Olaniyan, Olugbemi T., Adetunji, Charles O., Adeniyi, Mayowa J., & Hefft, Daniel Ingo (2022b). Computational intelligence in IoT healthcare. In *Deep learning, machine learning and IoT in biomedical and health informatics* (1st ed.). First Published 2022. Imprint CRC Press. Pages 13. eBook ISBN 9780367548445. DOI: 10.1201/9780367548445-19

Ong, E. Z., Chan, Y. F. Z., Leong, W. Y., Lee, N. M. Y., Kalimuddin, S., Mohideen, S. M. H., Chan, K. S., Tan, A. T., Bertoletti, A., & Ooi, E. E. (2020). A dynamic immune response shapes COVID-19 progression. *Cell Host & Microbe*, *27*(6), 879–882.

Osman, M., Faridi, R. M., Sligl, W., Shabani-Rad, M.-T., Dharmani-Khan, P., Parker, A., Kalra, A., Tripathi, M. B., Storek, J., & Cohen Tervaert, J. W. (2020). Impaired natural killer cell counts and cytolytic activity in patients with severe COVID-19. *Blood Advances*, *4*(20), 5035–5039.

Ouchicha, C., Ammor, O., & Meknassi, M. (2020). CVDNet: A novel deep learning architecture for detection of coronavirus (Covid-19) from chest x-ray images. *Chaos, Solitons & Fractals*, *140*, 110245.

Oyedara, Omotayo Opemipo, Adeyemi, Folasade Muibat, Adetunji, Charles Oluwaseun, & Elufisan, Temidayo Oluyomi (2022). Repositioning antiviral drugs as a rapid and cost-effective approach to discover treatment against SARS-CoV-2 infection. In *Medical biotechnology, biopharmaceutics, forensic science and bioinformatics* (1st ed.). First Published 2022. Imprint CRC Press. Pages 12. eBook ISBN 9781003178903. DOI: 10.1201/9781003178903-10

Parbat, D., & Chakraborty, M. (2020). A python based support vector regression model for prediction of COVID19 cases in India. *Chaos, Solitons & Fractals*, *138*, 109942.

Peng, M., Yang, J., Shi, Q., Ying, L., Zhu, H., Zhu, G., Ding, X., He, Z., Qin, J., & Wang, J. (2020). *Artificial intelligence application in COVID-19 diagnosis and prediction.*

Pham, Q.-V., Nguyen, D. C., Huynh-The, T., Hwang, W.-J., & Pathirana, P. N. (2020). Artificial intelligence (AI) and big data for coronavirus (COVID-19) pandemic: A survey on the state-of-the-arts. *IEEE Access*, *8*, 130820.

Pirouz, B., Golmohammadi, A., Saeidpour Masouleh, H., De Lazzari, C., Violini, G., & Pirouz, B. (2020). Relationship between average daily temperature and average cumulative daily rate of confirmed cases of COVID-19. *Medrxiv*, 2020–04.

Polsinelli, M., Cinque, L., & Placidi, G. (2020). A light CNN for detecting COVID-19 from CT scans of the chest. *Pattern Recognition Letters*, *140*, 95–100.

Rahmatizadeh, S., Valizadeh-Haghi, S., & Dabbagh, A. (2020). The role of artificial intelligence in management of critical COVID-19 patients. *Journal of Cellular and Molecular Anesthesia*, *5*(1), 16–22.

Rasheed, J., Jamil, A., Hameed, A. A., Aftab, U., Aftab, J., Shah, S. A., & Draheim, D. (2020). A survey on artificial intelligence approaches in supporting frontline workers and decision makers for the COVID-19 pandemic. *Chaos, Solitons & Fractals*, *141*, 110337.

Raza, K. (2020). Artificial intelligence against COVID-19: A meta-analysis of current research. In A.-E. Hassanien, N. Dey, & S. Elghamrawy (Eds.), *Big Data Analytics and Artificial Intelligence against COVID-19: Innovation Vision and Approach* (pp. 165–176). Springer International Publishing. https://doi.org/10.1007/978-3-030-55258-9_10

Rica, S., & Ruz, G. A. (2020). Estimating SIR model parameters from data using differential evolution: An application with COVID-19 data. *2020 IEEE Conference on Computational Intelligence in Bioinformatics and Computational Biology (CIBCB)*, 1–6. https://ieeexplore.ieee.org/abstract/document/9277708/

Richardson, S. J., Carroll, C. B., Close, J., Gordon, A. L., O'Brien, J., Quinn, T. J., Rochester, L., Sayer, A. A., Shenkin, S. D., & van der Velde, N. (2020). Research with older people in a world with COVID-19: Identification of current and future priorities, challenges and opportunities. *Age and Ageing*, *49*(6), 901–906.

Salman, F. M., Abu-Naser, S. S., Alajrami, E., Abu-Nasser, B. S., & Alashqar, B. A. (2020). Covid-19 detection using artificial intelligence. *IJAER*, *4*, 18–25.

Schuller, B. W., Batliner, A., Bergler, C., Mascolo, C., Han, J., Lefter, I., Kaya, H., Amiriparian, S., Baird, A., Stappen, L., Ottl, S., Gerczuk, M., Tzirakis, P., Brown, C., Chauhan, J., Grammenos, A., Hasthanasombat, A., Spathis, D., Xia, T., … Kaandorp, C. (2021). *The INTERSPEECH 2021 Computational Paralinguistics Challenge: COVID-19 Cough, COVID-19 Speech, Escalation & Primates* (arXiv:2102.13468). arXiv. http://arxiv.org/abs/2102.13468

Schultz, M. B., Vera, D., & Sinclair, D. A. (2020). Can artificial intelligence identify effective COVID-19 therapies? *EMBO Molecular Medicine*, *12*(8), e12817.

Senthilraja, M. (2021). Application of Artificial Intelligence to Address Issues Related to the COVID-19 Virus. *SLAS Technology*, *26*(2), 123–126. https://doi.org/10.1177/2472630320983813

Shi, F., Wang, J., Shi, J., Wu, Z., Wang, Q., Tang, Z., He, K., Shi, Y., & Shen, D. (2020). Review of artificial intelligence techniques in imaging data acquisition, segmentation, and diagnosis for COVID-19. *IEEE Reviews in Biomedical Engineering*, *14*, 4–15.

Shrestha, R., Shrestha, S., Khanal, P., & Kc, B. (2020). Nepal's first case of COVID-19 and public health response. *Journal of Travel Medicine*, *27*(3), taaa024.

Stebbing, J., Krishnan, V., de Bono, S., Ottaviani, S., Casalini, G., Richardson, P. J., Monteil, V., Lauschke, V. M., Mirazimi, A., & Youhanna, S. (2020). Mechanism of baricitinib supports artificial intelligence-predicted testing in COVID-19 patients. *EMBO Molecular Medicine*, *12*(8), e12697.

Sujath, R., Chatterjee, J. M., & Hassanien, A. E. (2020). A machine learning forecasting model for COVID-19 pandemic in India. *Stochastic Environmental Research and Risk Assessment: Research Journal*, *34*(7), 959–972. https://doi.org/10.1007/s00477-020-01827-8

Summers, R. M. (2021). Artificial intelligence of COVID-19 imaging: A hammer in search of a nail. In *Radiology* (Vol. 298, Issue 3, pp. E162–E164). Radiological Society of North America.

Swayamsiddha, S., Prashant, K., Shaw, D., & Mohanty, C. (2021). The prospective of Artificial Intelligence in COVID-19 Pandemic. *Health and Technology*, *11*(6), 1311–1320. https://doi.org/10.1007/s12553-021-00601-2

Tamang, S. K., Singh, P. D., & Datta, B. (2020). Forecasting of Covid-19 cases based on prediction using artificial neural network curve fitting technique. *Global Journal of Environmental Science and Management*, *6*(Special Issue (Covid-19)), 53–64.

Tang, F., Liang, J., Zhang, H., Kelifa, M. M., He, Q., & Wang, P. (2021). COVID-19 related depression and anxiety among quarantined respondents. *Psychology & Health*, *36*(2), 164–178. https://doi.org/10.1080/08870446.2020.1782410

Tayarani, M. (2020). Applications of artificial intelligence in battling against covid-19: A literature review. *Chaos, Solitons & Fractals*.

Thomas, S. P., Fraum, T. J., Ngo, L., Harris, R., Balesh, E., Bashir, M. R., & Wildman-Tobriner, B. (2022). Leveraging artificial intelligence to enhance peer review: Missed liver lesions on computed tomographic pulmonary angiography. *Journal of the American College of Radiology*, *19*(11), 1286–1294.

Tiwari, A. (2020). Modelling and analysis of COVID-19 epidemic in India. *Journal of Safety Science and Resilience*, *1*(2), 135–140.

Tomar, A., & Gupta, N. (2020). Prediction for the spread of COVID-19 in India and effectiveness of preventive measures. *Science of The Total Environment*, *728*, 138762. https://doi.org/10.1016/j.scitotenv.2020.138762

Trott, O., & Olson, A. J. (2010). AutoDock Vina: Improving the speed and accuracy of docking with a new scoring function, efficient optimization and multithreading. *Journal of Computational Chemistry*, *31*(2), 455–461. https://doi.org/10.1002/jcc.21334

Vaid, S., Kalantar, R., & Bhandari, M. (2020). Deep learning COVID-19 detection bias: Accuracy through artificial intelligence. *International Orthopaedics*, *44*(8), 1539–1542.

Vaishya, R., Javaid, M., Khan, I. H., & Haleem, A. (2020). Artificial Intelligence (AI) applications for COVID-19 pandemic. *Diabetes & Metabolic Syndrome: Clinical Research & Reviews*, *14*(4), 337–339.

Wang, Y., Wang, Y., Chen, Y., & Qin, Q. (2020). Unique epidemiological and clinical features of the emerging 2019 novel coronavirus pneumonia (COVID-19) implicate special control measures. *Journal of Medical Virology*, *92*(6), 568–576. https://doi.org/10.1002/jmv.25748

Wu, J. T., Leung, K., Bushman, M., Kishore, N., Niehus, R., de Salazar, P. M., Cowling, B. J., Lipsitch, M., & Leung, G. M. (2020). Estimating clinical severity of COVID-19 from the transmission dynamics in Wuhan, China. *Nature Medicine*, *26*(4), 506–510.

Wynants, L., Van Calster, B., Collins, G. S., Riley, R. D., Heinze, G., Schuit, E., Albu, E., Arshi, B., Bellou, V., & Bonten, M. M. (2020). Prediction models for diagnosis and prognosis of covid-19: Systematic review and critical appraisal. *BMJ*, *369*. https://www.bmj.com/content/369/bmj.m1328.abstract

Xu, L., Mao, Y., & Chen, G. (2020). Risk factors for 2019 novel coronavirus disease (COVID-19) patients progressing to critical illness: A systematic review and meta-analysis. *Aging (Albany NY)*, *12*(12), 12410.

Xu, Y., Liu, X., Cao, X., Huang, C., Liu, E., Qian, S., Liu, X., Wu, Y., Dong, F., & Qiu, C.-W. (2021). Artificial intelligence: A powerful paradigm for scientific research. *The Innovation*, *2*(4). https://www.cell.com/article/S2666-6758(21)00104-1/fulltext

Yahya, M., Maftuhati, M., Mustofa, A. H., & Arifa, Z. (2021). Online-based Arabic learning management during the Covid-19 pandemic era: Plan, implementation and evaluation. *Al-Ta'rib: Jurnal Ilmiah Program Studi Pendidikan Bahasa Arab IAIN Palangka Raya*, *9*(1), 85–98.

Yang, Y., Lu, Q.-B., Liu, M.-J., Wang, Y.-X., Zhang, A.-R., Jalali, N., Dean, N. E., Longini, I., Halloran, M. E., & Xu, B. (2020). Epidemiological and clinical features of the 2019 novel coronavirus outbreak in China. *Medrxiv*, https://doi.org/10.1101/2020.02.10.20021675

Zen, M., Fuzzi, E., Astorri, D., Saccon, F., Padoan, R., Ienna, L., Cozzi, G., Depascale, R., Zanatta, E., Gasparotto, M., Benvenuti, F., Bindoli, S., Gatto, M., Felicetti, M., Ortolan, A., Campaniello, D., Larosa, M., Lorenzin, M., Ramonda, R., … Doria, A. (2020). SARS-CoV-2 infection in patients with autoimmune rheumatic diseases in northeast Italy: A cross-sectional study on 916 patients. *Journal of Autoimmunity*, *112*, 102502. https://doi.org/10.1016/j.jaut.2020.102502

Zhang, J., Gao, X., Kan, J., Ge, Z., Han, L., Lu, S., Tian, N., Lin, S., Lu, Q., Wu, X., Li, Q., Liu, Z., Chen, Y., Qian, X., Wang, J., Chai, D., Chen, C., Li, X., Gogas, B. D., … Chen, S.-L. (2018). Intravascular Ultrasound Versus Angiography-Guided Drug-Eluting Stent Implantation. *Journal of the American College of Cardiology*, *72*(24), 3126–3137. https://doi.org/10.1016/j.jacc.2018.09.013

Zhang, J., Lu, H., Zeng, H., Zhang, S., Du, Q., Jiang, T., & Du, B. (2020). The differential psychological distress of populations affected by the COVID-19 pandemic. *Brain, Behavior, and Immunity*, *87*, 49–50. https://doi.org/10.1016/j.bbi.2020.04.031

Zhou, H., Chen, X., Hu, T., Li, J., Song, H., Liu, Y., Wang, P., Liu, D., Yang, J., Holmes, E. C., Hughes, A. C., Bi, Y., & Shi, W. (2020a). A Novel Bat Coronavirus Closely Related to SARS-CoV-2 Contains Natural Insertions at the S1/S2 Cleavage Site of the Spike Protein. *Current Biology*, *30*(11), 2196–2203. https://doi.org/10.1016/j.cub.2020.05.023

Zhou, Y., Wang, F., Tang, J., Nussinov, R., & Cheng, F. (2020b). Artificial intelligence in COVID-19 drug repurposing. *The Lancet Digital Health*, *2*(12), e667–e676.

3 Emerging Technologies That Are Relevant in Protecting Health Data and Easy Communication Between Patients and Healthcare Providers

Olorunsola Adeyomoye
University of Medical Sciences, Ondo City, Nigeria

Charles Oluwaseun Adetunji
Edo State University, Uzairue, Nigeria

Olugbemi T. Olaniyan
Kwara state university, Ilorin Kwara State

Juliana Bunmi Adetunji and Akinola Samson Olayinka
Edo State University, Uzairue, Nigeria

Oluwafemi Adebayo Oyewole
Federal University of Technology, Minna, Nigeria

K. I. T. Eniola
Joseph Ayo Babalola University, Ikeji Arakeji, Nigeria

DOI: 10.1201/9781003309468-3

3.1 INTRODUCTION

Rapid and effective diagnostic technologies have continued to improve global healthcare through the development of new vaccines, effective drug delivery mechanisms, and innovative therapies in addition public health measures (Häneke & Sahara, 2022). The introduction of three biotechnologies, including DNA recombinant techniques, monoclonal antibody technology, and the development of microchemical instrumentation, has allowed rapid and effective integration and sequence analysis of proteins and genes, and has changed biotechnology and medicine over the last decade (Boissel et al., 2015). The introduction of genetic engineering, or integrated DNA-technology or genetic engineering, has been one of the most revolutionary advances in biology in recent years. Recent research has led to a better understanding of the basic pathophysiology of many single genetic disorders, genetic identification in circumstances where the biochemistry is unknown, and with valuable information about the pathophysiology of many common diseases (Børresen, 1989).

Recombinant DNA technology provides the scientist with the tools needed to test human genetic mutations at the DNA sequence level and to determine how the mutant genes influence genetic expression. The contribution of science to improving public health and reducing global health inequality is more crucial than ever, given the great potential of genomics and other advances in health sciences (Zhong et al., 2021). However, millions of individuals in developing nations have yet to benefit from modern treatment (Acharya et al., 2004). It is vital to remember that science and technology can be used more successfully in developing nations in association with public health systems to improve their effectiveness. Health informatics evolved as a structured concept over time, with origins in the history of technological knowledge and medicine. Their development continues, and today's labor will become the history of the future. Despite excellent intentions, the history of health informatics is littered with promises of ideas that never materialized. Experience with the use of information technologies to improve healthcare often gives the appropriate perspective for individuals studying informatics (Cesnik & Kidd, 2010).

Despite numerous initiatives and apparent benefits of exchanging patient data with healthcare providers, government bodies have continued to raise concerns about exchanging of health information (Medford-Davis et al., 2017). The government's effective delivery of health information has been hampered by difficulties in ensuring international cooperation and security concerns. For ages, governments have given healthcare as a basic service to their citizens. The quality of health services supplied by health organizations and experts has changed dramatically during the last few decades (Abdulnabi et al., 2017; Adetunji et al., 2020; Adetunji et al., 2020; Oyedara et al., 2022; Adetunji et al., 2022a, b, c, d, e, f, g, h; Olaniyan et al., 2022a).

Electronic health has emerged as a result of recent advancements, making it more possible through the dissemination and adoption of information and communication technology. Cybercriminals and attackers, on the other hand, take advantage of ICT-related flaws, resulting in data breaches involving patients' personal health information (Zeadally et al., 2016).

3.2 INNOVATIVE BIOTECHNOLOGIES IN HEALTHCARE SYSTEM

Biocompatible materials, seed cells, and supporting components can now be 3D printed into active 3D living tissue, thanks to recent breakthroughs. Furthermore, 3D bioprinting is propelling key advancements in regenerative medicine (Ma et al., 2019). The scaffolding with 3D printing is commonly used in clinical studies and on the laboratory bench. Tissue bioprinting with living cells and matrix material is still a work in progress.

Nanomedicine advances include the application of nanobiotechnologies in medicine. It encompasses advances in molecular diagnostics, nanopharmaceutics, drug discovery, design, and delivery. Many novel medical and surgical equipment, such as nanorobots, are based on nanobiotechnology (Jain, 2008). It applies to almost all medical fields and the emerging examples of those related to cancer (nanooncology), nanoneurology, cardiovascular disorders (nanocardiology), orthopedics, nanoophthalmology, and infectious diseases.

Nanobiotechnology facilitates the integration of diagnostic and therapeutic drugs and facilitate the development of personalized drugs, i.e. the provision of specific treatments that are most appropriate for each individual. Many changes have already occurred, and within a decade, the medical profession will see a significant change. Tandem mass spectrometry development has made possible applications in numerous domains of medicine and biotechnology, including novel therapeutic preparation, manufacturing of active biological molecules, identification of pathogenic microbes, and causative agents of extremely pathogenic diseases. Furthermore, microelectromechanical systems are technologies developed from the integrated circuit industry to build sensors and actuators. To understand the design of MEMS equipment, MEMS design incorporates overhead, mass micromachining, and molding are essential. It has several medical and biological functions such as pressure sensors, accelerometers, human retinal prosthesis, affected sensory skin and MEMS-based kidney restoration system (Panescu, 2006). Blockchain technology is another inventive force at work in a range of industries, and its potential to change business operations has piqued interest (Chang & Chen, 2020). To assess the dynamic patterns of blockchain-based healthcare implementation, researchers looked at the stages of birth, expansion, leadership, renewal, and death in the life cycle of the business ecosystem.

Blockchain not only improves player contact but also the establishment of an ecosystem life cycle. Within the healthcare community, the blockchain-linked network could play an essential role in generating value, transferring information, and sharing data. Although new and developing technologies can serve to improve the quality of healthcare, implementing them in a perioperative setting can present a number of problems (Catalano & Fickenscher, 2007). Employees should be informed of future healthcare domains as well as the types of technological advancements they might anticipate. Some of the developing technologies that are transforming the perioperative environment include robots, digital displays, patient tracking systems, artificial limbs, and magnetic sensors (Francis, 2018). Employees should be ready to implement solutions that address the practical issues of incorporating this new technology into the perioperative setting. Artificial intelligence (AI) is a term that refers to the use of a computer to simulate intelligent behavior with little or no human interaction (Hamet & Tremblay, 2017).

The use of AI is widely regarded as the beginning of robot development. The word encompasses a wide range of medical fields, including medical diagnostics, robotics, medical statistics, and human biology, as well as recent omics technologies (Chen & Decary, 2020). The visual branch includes informatics approaches such as in-depth learning information management, health management systems, such as electronic health records, and effective physician guidance in treatment decisions (Pashkov et al., 2020). Robots that aid an elderly patient or surgeon are the best example of the physical branch. Targeted nanorobots, a new and different medication delivery technology, are also included in this field. In the current phase of medical AI development it exists in three technologies: software, hardware, and mixed forms using three main mathematical methods, i.e. the flow chart method, the data storage method, and the decision-making method. All of them are usable but uniquely suitable for AI implementation (Jambaulikar et al., 2021). The main problems of implementing AI in healthcare are related to the nature of the technology itself, the difficulty of legal support regarding safety and efficiency, privacy, ethics and credit concerns.

3.3 APPLICATION OF BIOTECHNOLOGIES IN MEDICINE

Nanomedicine is the medical application of nanotechnology, and it is widely employed in the healthcare industry. Doctors, scientists, and specialists have used nanotechnology to better understand cellular changes in order to produce nanomedicines and address healthcare concerns (Ravindran et al., 2018). Nanoparticles with a diameter of less than 1 nm are employed as medication delivery and genetic delivery systems in humans to speed up therapeutic action. Nanomaterials are similar in size to biomolecules and are predicted to interact more effectively (Jain, 2010).

Understanding the potency, toxicity, and safe usage of nanoparticles requires knowledge of their pharmacokinetics, metabolism, permeability, distribution, and termination. Toxicological studies on nanoparticles have yielded results, but knowledge of metabolism, pharmacokinetics, distribution, and approval of nanomedicine is lacking (Patra et al., 2018). As a result, information on nanomaterial metabolism, pharmacokinetics, permeability, and biodistribution is useful in nanomedicine. Nanomedicine is becoming increasingly significant in illness treatment and diagnostics. The size of the particles is crucial. The therapeutic impact increases as the particle size decreases. Nanomedicine has been reported to have improved pharmacokinetics, bioavailability, half-life, metabolism, biodistribution, and permeability than bigger medicines. In organic medicine, in vitro and in vivo ADME research are required. Nanomatadium should also be included in ADME research, both in vitro and in vivo. As a result, nanomedicine can aid in the production of safe human personal drugs. The use of nanoscale devices as a diagnostic tool or to administer therapeutic substances to particular targeted areas in a controlled manner is relatively new but quickly evolving in nanomedicine and nano delivery systems (Xuan et al., 2013). Nanotechnology has a number of applications in the treatment of certain human diseases and the targeted administration of specific medications. There have been numerous notable applications of nanomedicine in the treatment of various diseases recently, including chemotherapeutic medicines, biological agents, immunotherapeutic agents, and so on.

Recent advances in the field of nanomedicines and nano-based drug delivery systems, as well as special diagnostics using disease markers, have been made through comprehensive analysis of the detection and use of nanomaterials in improving both the efficacy of both new and old drugs (e.g., natural products). Nanooncology, or the application of nanobiotechnology to cancer treatment, is the most important branch of nanomedicine right now. Using gold nanoparticles and quantum dots, nanobiotechnology has refined and increased the diagnostic limitations of cancer cells. The essential identification of multiple protein biomarkers by nanobiosensors is one example of how nanobiotechnology has enhanced cancer biomarker detection. Magnetic nanoparticles can catch tumor cells in the bloodstream, which can then be detected via photoacoustics. Nanoparticles allow for the targeted administration of cancer treatments, which improves efficacy and reduces adverse effects by lowering the number of anti-cancer drugs available.

The detection of cancer biomarkers has also improved thanks to nanobiotechnology. Nanoparticulate anti-cancer medications can break down other cellular barriers, allowing for a therapeutic focus while also protecting the surrounding tissues from the toxins. Through surgery, nanoparticle constructions enable the transmission of various forms of energy. The administration of cancer therapy medicines can be integrated with nanoparticle-based optical imaging of tumors and various agents to improve tumor detection by magnetic resonance imaging. Monoclonal antibody nanoparticle complexes are being studied for cancer diagnostics and tailored treatment administration.

Chemotherapeutic drugs based on nanoparticles are already on the market, and a few are in clinical trials. Cancer self-treatment is based on a better understanding of the disease at the cellular level, which nanobiotechnology facilitates. Genome-scale metabolic models have become a prominent tool in systemic biology, with applications in sectors as diverse as industrial biotechnology and system medicine (Zhang & Hua, 2016). GEMs have recently attracted a lot of interest since more studies have been undertaken employing them.

The introduction of next-generation sequencing technology (NGS) has revolutionized the genomics business, allowing for the production of genetic sequence data with more precision and accuracy at a lower cost. Rapid technological advancements, led by educational institutions and businesses, have continued to expand NGS applications from research to clinics throughout the years. The medicinal significance of NGS technology on Mendelian and complicated diseases, particularly cancer, has recently been highlighted. However, the growing speed of NGS capture creates major obstacles in terms of data processing, storage, management, interpretation, and quality control, obstructing the translation of data sequencing to clinical performance (Nagashima et al., 2012). To combat this deficiency, scientists have considered the possibility of using animals as an organ. Xenografting is the process of transplanting organisms between different species. Selection of source animals will be crucial for xenotransplantation. Ideally, a donating animal can show immunological, anatomical, and physiological similarity in humans, have organs of the same size as humans, and can have a shorter generation.

3.4 INFORMATION SECURITY IN HEALTHCARE

Doctors, nurses, and other medical professionals spend a lot of time in front of computers, using software for general practitioners, specialty care, or an integrated hospital system, among other things (Newell & Jordan, 2015). Because the information they collect during recovery and patient care is frequently sensitive, their management is highly regulated. Current ways of accessing patient data, accumulation, donation, exchange, and control could be disrupted by blockchain technology. Patients can securely exchange data with providers and restrict access by utilizing interoperability standards, smart contracts, and encrypted identification (Hylock & Zeng, 2019). Long-term medical records, as a result, can drastically reduce the cost and quality of patient treatment for both individuals and groups. For patient-centered access and exchange of health data, blockchain is a powerful and effective tool. The health chain gives patients and clinicians access to consistent and full medical information by integrating systematic, collaborative, and patient-generated data into a globally accessible blockchain and sharing it via intelligent contracts.

Data security, interoperability, block storage, and access to patient-controlled data are all issues that must be addressed, as well as a few developing contexts that require more speed and safety considerations (Ködmön & Csajbók, 2015). There is a need to address a comprehensive solution for the protection of medical information systems: technical, procedural and legal protection of medical records, information, applications, hardware and software, networks and system management. When utilized to interact and share data between healthcare providers, patients, and researchers, the internet has several advantages. However, the advantages of the internet come at a far higher cost in terms of secrecy, integrity, and information availability (Ilioudis & Pangalos, 2001).

3.5 ARTIFICIAL INTELLIGENCE FOR SECURITY OF HEALTH INFORMATION EXCHANGE

Due to the high demand for healthcare services and the increasing capabilities of artificial intelligence, discussion agents have been developed to assist with a variety of health-related activities, such as behavioral change, medical support, health monitoring, training, evaluation, and testing (Milne-Ives et al., 2020). The automation of these activities can free physicians from focusing on strenuous work and increasing access to public healthcare services. To assemble evidence for future development, extensive evaluation of the acceptance, usefulness, and effectiveness of these healthcare agents is required. This will allow future development to discover areas for development and sustainable acquisition potential (Schachner et al., 2020). There are an increasing number of chat agents or chatbots that are equipped with artificial intelligence systems. There are an increasing variety of healthcare applications available, such as those that provide information and assistance to those suffering from chronic conditions. Chatbots powered by artificial intelligence provide efficient and common communication with such patients. AI enabled by chatbots can expand into genuine dialogues and build relationships with users (Zhang et al., 2020).

One of the promising avenues for developing low-cost and cheap behavioral interventions to enhance physical activity and good eating is the use of AI chatbots in lifestyle modification programs. Artificial intelligence is becoming more prevalent in the medical field. AI-powered chatbot systems can be used as automated chat agents to promote health, provide education, and change potential habits. To forecast detection, it's vital to assess the motivation for using health chatbots; nevertheless, only a few research have done thus far (Nadarzynski et al., 2019). The rise in demand for mental health treatments, combined with artificial intelligence's increasing capabilities, has accelerated the development of digital mental health solutions in recent years. DMHI has so far integrated AI-based chatbots to help with diagnostics and testing, symptom management and behavioral change, and content distribution (Nadarzynski et al., 2019).

3.6 BIG DATA IN MEDICINE AND HEALTHCARE DELIVERY

Machine learning-based big data analysis has a lot of advantages when it comes to comparing and assessing massive amounts of complex healthcare data. However, in order to effectively apply machine learning techniques in healthcare, a number of constraints must be addressed, as well as crucial problems such as clinical application and ethical standards in healthcare delivery (Ngiam & Khor, 2019). Machine learning has more flexibility than traditional biostatistical methods, making it helpful for a variety of tasks like risk classification, diagnostics and classification, and survival predicting. Machine learning algorithms also have the ability to assess many sorts of data (for example, statistical data, laboratory findings, photographic data, and free medical notes) and incorporate predictors of illness risk, diagnosis, prognosis, and suitable treatment (Adetunji et al., 2021).

Despite these advantages, applying machine learning to healthcare has unique issues that necessitate prior data analysis, modeling training, and system improvement in connection to the actual clinical situation. Model Health is a platform that attempts to make machine learning tactics in medical data easier to deploy in order to improve healthcare delivery (Pitoglou et al., 2019). In many aspects, including neural networks, the field of healthcare is a full manner of adopting procedures for the creation and implementation of machine learning algorithms, and can be used to aid clinical practice and management decisions (Hu, 2021). It covers the entire life cycle of these operations, from initial data pumping, duplication, anonymization, and enrichment to ultimate disposal of algorithms that are compatible with Application Program Interfaces and may be used by any approved Information system. Recently, there has been a surge in interest in designing technologies to help oncologists make more effective and reliable treatment decisions (Naqa et al., 2018). This has significantly fueled the demand for more personalized and precise oncology in the so-called big data era, which promises to change cancer therapy by harnessing the power of huge data flows. This surge in interest in big data statistics has brought with it new opportunities and issues that have yet to be addressed. These include standard clinical data integration and suspension; patient privacy; modification of current analytical methods for dealing with such noisy and distinct data; and expanded use of advanced mathematical learning methods based on the integration of modern mathematical and mechanical learning methods (Mooney & Pejaver, 2018).

The digital world generates data at an incredible and ever-increasing rate. Despite the fact that "big data" has opened up new avenues for understanding public health, it still has a lot of research and practice potential. Medical renaming is now commonly referred as "big data." The rapid advancement of machine learning techniques and practical abilities, in particular, has the potential to revolutionize medical practice, from resource allocation to difficult disease detection (Price & Cohen, 2019). However, big data comes with a slew of risks and obstacles, including serious concerns about patient privacy. When developing patient-centered health technology, it's critical to think about privacy and the consequences of big data policy (Thakkar & Gordon, 2019). The integration and analysis of large complex data, such as data on various omics (genomics, epigenomics, transcriptomics, proteomics, metabolomics, internomics, pharmacogenomics, and diseasomics), biomedical data, and electronic health records data, is required for large-scale medical and healthcare data analysis (Ristevski & Chen, 2018). Privacy and security challenges with big data are difficult to resolve. Additional advice for choosing the right and promising distributed data processing software platform are provided for massive data features.

3.7 ENHANCING HEALTHCARE DELIVERY THROUGH EFFECTIVE HEALTH INFORMATION EXCHANGE

Patients experiencing behavioral health and chronic medical conditions often overuse the hospital services of inpatients. This overuse pattern contributes to the use of inappropriate healthcare (Parker et al., 2016). These patients need integrated care to achieve optimal health outcomes. However, improper exchange of health-related information between different nurses poses a challenge to the delivery of integrated care (Martin et al., 2018). The health information exchange (HIE) facilitates the sharing of health-related information and has been shown to be effective in the management of chronic diseases; however, their effectiveness in the delivery of integrated care is unclear. It is wise to consider new ways to share common and ethical health information. According to the SAFR model search, warning, file, and reconciliation created in partnership with emergency medical services, health information agencies have defined the benefits of exchanging health information for emergency medical services (Hersh et al., 2015). HIE use has risen over time, with hospital use being much higher and long-term care settings being significantly lower. However, HIE is still underutilized in donor organizations. A lack of active interactive marketing, inert workflow, and a poorly designed visual interface and review tools are all barriers to using HIE.

Many promoters and barriers to implementation and sustainability have been found through research, but no studies have yet been conducted to assess or compare their impact (Nakayama et al., 2021). Future research should address a wide range of topics, use rigorous designs, use a standard definition of HIE types, and be part of a systematic approach to learning HIE in order to improve our understanding of HIE. By providing clinicians with competent guidance, sharing health information can increase diagnosis accuracy, effective treatment, and safety. Many past HIE investigations, on the other hand, focused on nature (Sadoughi et al., 2018). HIE is a well-known technology that allows all clinical and administrative data to be shared electronically across all healthcare settings. Despite the fact that this technology has tremendous potential in the healthcare business, there is less proof of its impact on care quality and cost effectiveness.

3.8 CURRENT CHALLENGES AND FUTURE PERSPECTIVES

Among the concerns surrounding healthcare systems, technological integration is a prominent topic. Several papers show how computing can be used wherever to improve the efficiency of healthcare delivery (Bott et al., 2005). However, because of the technology's surrounding but invisible nature, computing everywhere is also a social challenge. To transform the ideas of this exciting new research into reality, contributions from practically all of the lower established medical informatics fields are required. In recent years, artificial intelligence technology has piqued the interest of a wide range of industries, including health (Guimarães, 2014). The growth of computer hardware and software applications in medicine, as well as the digital integration of health-related data, all contribute to progress in the creation and implementation of new technologies. Methods that use a multi-sectoral biomedical informatics to incorporate domain knowledge might improve the functionality of modern AI machine learning algorithms and help them overcome problems like scarce, missing, or inconsistent data.

New heuristics for training and encryption techniques may make it easier for people to learn about secrecy (Wang & Preininger, 2019). To solve cost concerns, healthcare organizations around the world are heavily investing in value-based healthcare, and time-based, work-based prices have been raised as part of the cost. It is believed that problem-based learning approaches and disciplinary collaboration are crucial to solving the present problems of health information development and implementation in the healthcare system (Jackson et al., 2019). In order to distribute health information at the full level, the triple aim of providing better healthcare at a cheaper cost to the community is connected with the same goals at the institutional level and to the person. We must embrace concepts like collaborative innovation and creative collaboration with users and actors in order to reach this goal (Keel et al., 2017). As indicated by its extensive use and quick growth, electronic communication is becoming increasingly popular around the world. However, it is still a relatively new approach to reach patients in medicine. They do, however, have the opportunity to improve present healthcare. Treatment adherence and communication between physicians and patients can be aided through electronic platforms. Furthermore, a systematic review of the Screome social media platform's content has identified health events that aren't usually taken for granted by general healthcare by better identifying these healthcare events (Botin et al., 2015). As we enhance our ability to match the demands of patients, strategies in our current healthcare system can be improved. All of these tactics are critical components of modern healthcare and will aid us in the development of digital healthcare in the future (Gijsen et al., 2020).

3.9 CONCLUSION

Nanotechnology has improved the ability to combine, manage, and visualize objects on a nanometer scale. New technological advances in nanoscale have brought an unprecedented opportunity to direct the scope of biomolecular interactions, as well as the motivation to create intelligent nanostructures for successful therapeutic approaches. The continuous integration of bio- and nanotechnologies is beginning to produce a revolution in diagnostic, therapeutic, and monitoring diseases and unresolved medical problems. The use of health information to prevent and improve information exchange has transformed health systems. A major challenge results in the lack of interoperability and standardization of interfaces of this system, which hinders effective collaboration and information sharing in complex patient care.

REFERENCES

Abdulnabi, M., Al-Haiqi, A., Kiah, M., Zaidan, A. A., Zaidan, B. B., & Hussain, M. (2017). A distributed framework for health information exchange using smartphone technologies. *Journal of Biomedical Informatics*, *69*, 230–250.

Acharya, T., Kennedy, R., Daar, A. S., & Singer, P. A. (2004). Biotechnology to improve health in developing countries -- A review. *Memórias do Instituto Oswaldo Cruz*, *99*(4), 341–350.

Adetunji, Charles Oluwaseun, Egbuna, Chukwuebuka, Oladosun, Tolulope Olawumi, Akram, Muhammad, Michael, Olugbenga, Olisaka, Frances Ngozi, Ozolua, Phebean, Adetunji, Juliana Bunmi, Enoyoze, Goddidit Esiro, & Olaniyan, Olugbemi (2021). Efficacy of phytochemicals of medicinal plants for the treatment of human echinococcosis. In *Neglected tropical diseases and phytochemicals in drug discovery*. DOI: 10.1002/9781119617143.ch8

Adetunji, Charles Oluwaseun, Mitembo, William Peter, Egbuna, Chukwuebuka, Narasimha Rao, G.M. (2020). In silico modeling as a tool to predict and characterize plant toxicity. In Andrew G. Mtewa, Chukwuebuka Egbuna, & G.M. Narasimha Rao (Eds.), *Poisonous plants and phytochemicals in drug discovery*. Wiley Online Library DOI: 10.1002/9781119650034.ch14

Adetunji, Charles Oluwaseun, Nwankwo, Wilson, Olayinka, Akinola Samson, Olugbemi, Olaniyan Tope, Akram, Muhammad, Laila, Umme, Olugbenga, Michael Samuel, Oshinjo, Ayomide Michael, Adetunji, Juliana Bunmi, Okotie, Gloria E., & Esiobu, Nwadiuto (Diuto) (2022h). Machine learning and behaviour modification for COVID-19. In *Medical biotechnology, biopharmaceutics, forensic science and bioinformatics* (1st ed.). First Published 2022. Imprint CRC Press. Pages 17. eBook ISBN 9781003178903. DOI: 10.1201/9781003178903-17

Adetunji, Charles Oluwaseun, Nwankwo, Wilson, Olayinka, Akinola Samson, Olugbemi, Olaniyan Tope, Akram, Muhammad, Laila, Umme, Samuel, Michael Olugbenga, Oshinjo, Ayomide Michael, Adetunji, Juliana Bunmi, Okotie, Gloria E., & Esiobu, Nwadiuto (Diuto) (2022f). Computational intelligence techniques for combating COVID-19. In *Medical biotechnology, biopharmaceutics, forensic science and bioinformatics* (1st ed.). First Published 2022. Imprint CRC Press. Pages 12. eBook ISBN 9781003178903. DOI: 10.1201/9781003178903-16

Adetunji, Charles Oluwaseun, Olaniyan, Olugbemi Tope, Adeyomoye, Olorunsola, Dare, Ayobami, Adeniyi, Mayowa J, Alex, Enoch, Rebezov, Maksim, Garipova, Larisa, & Shariati, Mohammad Ali. (2022a). eHealth, mHealth, and telemedicine for COVID-19 pandemic. In S.K. Pani, S. Dash, W.P. dos Santos, S.A. Chan Bukhari, & F. Flammini (Eds.), *Assessing COVID-19 and other pandemics and epidemics using computational modelling and data analysis*. Springer. DOI: 10.1007/978-3-030-79753-9_10

Adetunji, Charles Oluwaseun, Olaniyan, Olugbemi Tope, Adeyomoye, Olorunsola, Dare, Ayobami, Adeniyi, Mayowa J., Alex, Enoch, Rebezov, Maksim, Isabekova, Olga, & Shariati, Mohammad Ali (2022c). Smart sensing for COVID-19 pandemic. In S.K. Pani, S. Dash, W.P. dos Santos, S.A. Chan Bukhari, & F. Flammini (Eds.), *Assessing COVID-19 and other pandemics and epidemics using computational modelling and data analysis.* Springer. DOI: 10.1007/978-3-030-79753-9_9

Adetunji, Charles Oluwaseun, Olaniyan, Olugbemi Tope, Adeyomoye, Olorunsola, Dare, Ayobami, Adeniyi, Mayowa J., Alex, Enoch, Rebezov, Maksim, Koriagina, Natalia, & Shariati, Mohammad Ali (2022e). Diverse techniques applied for effective diagnosis of COVID-19. In S.K. Pani, S. Dash, W.P. dos Santos, S.A. Chan Bukhari, & F. Flammini (Eds.), *Assessing COVID-19 and other pandemics and epidemics using computational modelling and data analysis.* Springer. DOI: 10.1007/978-3-030-79753-9_3

Adetunji, Charles Oluwaseun, Olaniyan, Olugbemi Tope, Olorunsola Adeyomoye, Ayobami Dare, Mayowa J. Adeniyi, Enoch Alex, Maksim Rebezov, Ekaterina Petukhova, & Shariati, Mohammad Ali (2022b). Machine learning approaches for COVID-19 pandemic. In S.K. Pani, S. Dash, W.P. dos Santos, S.A. Chan Bukhari, & F. Flammini (Eds.), *Assessing COVID-19 and other pandemics and epidemics using computational modelling and data analysis.* Springer. DOI: 10.1007/978-3-030-79753-9_8

Adetunji, Charles Oluwaseun, Olaniyan, Olugbemi Tope, Adeyomoye, Olorunsola, Dare, Ayobami, Adeniyi, Mayowa J., Alex, Enoch, Rebezov, Maksim, Petukhova, Ekaterina, & Shariati, Mohammad Ali (2022d). Internet of Health Things (IoHT) for COVID-19. In S.K. Pani, S. Dash, W.P. dos Santos, S.A. Chan Bukhari, & F. Flammini (Eds.), *Assessing COVID-19 and other pandemics and epidemics using computational modelling and data analysis.* Springer. DOI: 10.1007/978-3-030-79753-9_5

Adetunji, Charles Oluwaseun, Olugbemi, Olaniyan Tope, Akram, Muhammad, Laila, Umme, Samuel, Michael Olugbenga, Oshinjo, Ayomide Michael, Adetunji, Juliana Bunmi, Okotie, Gloria E., Esiobu, Nwadiuto (Diuto), Oyedara, Omotayo Opemipo, & Adeyemi, Folasade Muibat (2022g). Application of computational and bioinformatics techniques in drug repurposing for effective development of potential drug candidate for the management of COVID-19. In *Medical biotechnology, biopharmaceutics, forensic science and bioinformatics* (1st ed.). First Published 2022, Imprint CRC Press. Pages 14. eBook ISBN 9781003178903. DOI: 10.1201/9781003178903-15

Adetunji, Charles Oluwaseun, & Oyeyemi, Oyetunde T. (2022). Antiprotozoal activity of some medicinal plants against entamoeba histolytica, the causative agent of amoebiasis. In *Medical biotechnology, biopharmaceutics, forensic science and bioinformatics* (1st ed.). First Published 2022. Imprint CRC Press. Pages 12. eBook ISBN 9781003178903. https://www.taylorfrancis.com/chapters/edit/10.1201/9781003178903-20/antiprotozoal-activity-medicinal-plants-entamoeba-histolytica-causative-agent-amoebiasis-charles-oluwaseun-adetunji-oyetunde-oyeyemi

Akram, Muhammad, Adetunji, Charles Oluwaseun, Egbuna, Chukwuebuka, Jabeen, Shaista, Olaniyan, Olugbemi, Ezeofor, Nebechi Jane, Anani, Osikemekha Anthony, Laila, Umme, Găman, Mihnea-Alexandru, Patrick-Iwuanyanwu, Kingsley, Ifemeje, Jonathan Chinenye, Chikwendu, Chukwudi Jude, Michael, Olisah Chinedu, & Rudrapal, Mithun (2021b). Dengue fever. In *Neglected tropical diseases and phytochemicals in drug discovery.* DOI: 10.1002/9781119617143.ch17

Akram, Muhammad, Mohiuddin, Ejaz, Adetunji, Charles Oluwaseun, Oladosun, Tolulope Olawumi, Ozolua, Phebean, Olisaka, Frances Ngozi, Egbuna, Chukwuebuka, Michael, Olugbenga, Adetunji, Juliana Bunmi, Hameed, Leena, Awuchi, Chinaza Godswill, Patrick-Iwuanyanwu, Kingsley, & Olaniyan, Olugbemi (2021a). Chapter 7: Prospects of phytochemicals for the treatment of helminthiasis. In *Neglected tropical diseases and phytochemicals in drug discovery* (1st ed.). Wiley. DOI: 10.1002/9781119617143.ch7

Boissel, J. P., Auffray, C., Noble, D., Hood, L., & Boissel, F. H. (2015). Bridging systems medicine and patient needs. *CPT: Pharmacometrics & Systems Pharmacology*, *4*(3), e00026.

Børresen, A. L. (1989). Genteknologi i klinisk medisin-teknikker og verktøy [Gene technology in clinical medicine--technics and tools]. *Tidsskrift for den Norske laegeforening: tidsskrift for praktisk medicin, ny raekke*, *109*(28), 2882–2887.

Botin, L., Bertelsen, P., & Nøhr, C. (2015). Challenges in improving health care by use of health informatics technology. *Studies in Health Technology and Informatics*, *215*, 3–13.

Bott, O. J., Ammenwerth, E., Brigl, B., Knaup, P., Lang, E., Pilgram, R., Pfeifer, B., Ruderich, F., Wolff, A. C., Haux, R., & Kulikowski, C. (2005). The challenge of ubiquitous computing in health care: Technology, concepts and solutions. Findings from the IMIA Yearbook of Medical Informatics 2005. *Methods of Information in Medicine*, *44*(3), 473–479.

Catalano, K., & Fickenscher, K. (2007). Emerging technologies in the OR and their effect on perioperative professionals. *AORN Journal*, *86*(6), 958–969.

Cesnik, B., & Kidd, M. R. (2010). History of health informatics: A global perspective. *Studies in Health Technology and Informatics*, *151*, 3–8.

Chang, S. E., & Chen, Y. (2020). Blockchain in health care innovation: Literature review and case study from a business ecosystem perspective. *Journal of Medical Internet Research*, *22*(8), e19480.

Chen, M., & Decary, M. (2020). Artificial intelligence in healthcare: An essential guide for health leaders. *Healthcare Management Forum*, *33*(1), 10–18.

Francis, M. J. (2018). Recent advances in vaccine technologies. *The Veterinary Clinics of North America. Small Animal Practice*, *48*(2), 231–241.

Gijsen, V., Maddux, M., Lavertu, A., Gonzalez-Hernandez, G., Ram, N., Reeves, B., Robinson, T., Ziesenitz, V., Shakhnovich, V., & Altman, R. (2020). #Science: The potential and the challenges of utilizing social media and other electronic communication platforms in health care. *Clinical and Translational Science*, *13*(1), 26–30.

Guimarães R. (2014). Technological incorporation in the Unified Health System (SUS): the problem and ensuing challenges. *Ciência & Saúde Coletiva*, *19*(12), 4899–4908.

Hamet, P., & Tremblay, J. (2017). Artificial intelligence in medicine. *Metabolism, Clinical and Experimental*, *69S*, S36–S40.

Häneke, T., & Sahara, M. (2022). Progress in bioengineering strategies for heart regenerative medicine. *International Journal of Molecular Sciences*, *23*(7), 3482.

Hersh, W., Totten, A., Eden, K., Devine, B., Gorman, P., Kassakian, S., Woods, S. S., Daeges, M., Pappas, M., & McDonagh, M. S. (2015). Health information exchange. *Evidence Report/Technology Assessment*, (220), 1–465.

Hu M. (2021). Decision-making model of product modeling big data design scheme based on neural network optimized by genetic algorithm. *Computational Intelligence and Neuroscience*, *2021*, 9315700.

Hylock, R. H., & Zeng, X. (2019). A blockchain framework for patient-centered health records and exchange (HealthChain): Evaluation and proof-of-concept study. *Journal of Medical Internet Research*, *21*(8), e13592.

Ilioudis, C., & Pangalos, G. (2001). A framework for an institutional high level security policy for the processing of medical data and their transmission through the Internet. *Journal of Medical Internet Research*, *3*(2), E14.

Jackson, G., Hu, J., & Section Editors for the IMIA Yearbook Section on Artificial Intelligence in Health (2019). Artificial intelligence in health in 2018: New opportunities, challenges, and practical implications. *Yearbook of Medical Informatics*, *28*(1), 52–54.

Jain K. K. (2008). Nanomedicine: Application of nanobiotechnology in medical practice. *Medical Principles and Practice: International Journal of the Kuwait University, Health Science Centre*, *17*(2), 89–101.

Jain K. K. (2010). Advances in the field of nanooncology. *BMC Medicine*, *8*, 83.

Jambaulikar, G. D., Marshall, A., Hasdianda, M. A., Cao, C., Chen, P., Miyawaki, S., Baugh, C. W., Zhang, H., McCabe, J., Su, J., Landman, A. B., & Chai, P. R. (2021). Electronic paper displays in hospital operations: Proposal for deployment and implementation. *JMIR Formative Research*, *5*(8), e30862.

Keel, G., Savage, C., Rafiq, M., & Mazzocato, P. (2017). Time-driven activity-based costing in health care: A systematic review of the literature. *Health Policy (Amsterdam, Netherlands)*, *121*(7), 755–763.

Ködmön, J., & Csajbók, Z. E. (2015). Információbiztonság az egészségügyben [Information security in health care]. *Orvosi Hetilap*, *156*(27), 1075–1080.

Ma, Y., Xie, L., Yang, B., & Tian, W. (2019). Three-dimensional printing biotechnology for the regeneration of the tooth and tooth-supporting tissues. *Biotechnology and Bioengineering*, *116*(2), 452–468.

Martin, T. J., Ranney, M. L., Dorroh, J., Asselin, N., & Sarkar, I. N. (2018). Health information exchange in emergency medical services. *Applied Clinical Informatics*, *9*(4), 884–891.

Medford-Davis, L. N., Chang, L., & Rhodes, K. V. (2017). Health information exchange: What do patients want?. *Health Informatics Journal*, *23*(4), 268–278.

Milne-Ives, M., de Cock, C., Lim, E., Shehadeh, M. H., de Pennington, N., Mole, G., Normando, E., & Meinert, E. (2020). The effectiveness of artificial intelligence conversational agents in health care: Systematic review. *Journal of Medical Internet Research*, *22*(10), e20346.

Mooney, S. J., & Pejaver, V. (2018). Big data in public health: Terminology, machine learning, and privacy. *Annual Review of Public Health*, *39*, 95–112.

Nadarzynski, T., Miles, O., Cowie, A., & Ridge, D. (2019). Acceptability of artificial intelligence (AI)-led chatbot services in healthcare: A mixed-methods study. *Digital Health*, *5*, 2055207619871808.

Nagashima, H., Matsunari, H., Nakano, K., Watanabe, M., Umeyama, K., & Nagaya, M. (2012). Advancing pig cloning technologies towards application in regenerative medicine. *Reproduction in Domestic Animals = Zuchthygiene*, *47*Suppl 4, 120–126.

Nakayama, M., Inoue, R., Miyata, S., & Shimizu, H. (2021). Health information exchange between specialists and general practitioners benefits rural patients. *Applied Clinical Informatics*, *12*(3), 564–572.

Naqa, I. E., Kosorok, M. R., Jin, J., Mierzwa, M., & Ten Haken, R. K. (2018). Prospects and challenges for clinical decision support in the era of big data. *JCO Clinical Cancer Informatics*, *2*, CCI.18.00002.

Newell, S., & Jordan, Z. (2015). The patient experience of patient-centered communication with nurses in the hospital setting: a qualitative systematic review protocol. *JBI Database of Systematic Reviews and Implementation Reports*, *13*(1), 76–87.

Ngiam, K. Y., & Khor, I. W. (2019). Big data and machine learning algorithms for health-care delivery. *The Lancet. Oncology*, *20*(5), e262–e273.

Olaniyan, Olugbemi T., Adetunji, Charles O., Adeniyi, Mayowa J., & Hefft, Daniel Ingo (2022a). Machine learning techniques for high-performance Computing for IoT applications in healthcare. In *Deep learning, machine learning and IoT in biomedical and health informatics* (1st ed.). First Published 2022. Imprint CRC Press. Pages 13. eBook ISBN 9780367548445. DOI: 10.1201/9780367548445-20

Olaniyan, Olugbemi T., Adetunji, Charles O., Adeniyi, Mayowa J., & Hefft, Daniel Ingo (2022b). Computational intelligence in IoT healthcare. In *Deep learning, machine learning and IoT in biomedical and health informatics* (1st ed.). First Published 2022. Imprint CRC Press. Pages 13. eBook ISBN 9780367548445. DOI: 10.1201/9780367548445-19

Oyedara, Omotayo Opemipo, Adeyemi, Folasade Muibat, Adetunji, Charles Oluwaseun, & Elufisan, Temidayo Oluyomi (2022). Repositioning antiviral drugs as a rapid and cost-effective approach to discover treatment against SARS-CoV-2 infection. In *Medical biotechnology, biopharmaceutics, forensic science and bioinformatics* (1st ed.). First Published 2022. Imprint CRC Press. Pages 12. eBook ISBN 9781003178903. DOI: 10.1201/9781003178903-10

Panescu D. (2006). MEMS in medicine and biology. *IEEE Engineering in Medicine and Biology Magazine: The Quarterly Magazine of the Engineering in Medicine & Biology Society*, *25*(5), 19–28.

Parker, C., Weiner, M., & Reeves, M. (2016). Health information exchanges--Unfulfilled promise as a data source for clinical research. *International Journal of Medical Informatics*, *87*, 1–9.

Pashkov, V. M., Harkusha, A. O., & Harkusha, Y. O. (2020). Artificial intelligence in medical practice: regulative issues and perspectives. *Wiadomosci lekarskie (Warsaw, Poland: 1960)*, *73*(12 cz 2), 2722–2727.

Patra, J. K., Das, G., Fraceto, L. F., Campos, E., Rodriguez-Torres, M., Acosta-Torres, L. S., Diaz-Torres, L. A., Grillo, R., Swamy, M. K., Sharma, S., Habtemariam, S., & Shin, H. S. (2018). Nano based drug delivery systems: Recent developments and future prospects. *Journal of Nanobiotechnology*, *16*(1), 71.

Pitoglou, S., Anastasiou, A., Androutsou, T., Giannouli, D., Kostalas, E., Matsopoulos, G., & Koutsouris, D. (2019). MODEL health: Facilitating machine learning on big health data networks. In *Annual International Conference of the IEEE Engineering in Medicine and Biology Society. IEEE Engineering in Medicine and Biology Society. Annual International Conference, 2019* (pp. 2174–2177).

Price, W. N., & Cohen, I. G. (2019). Privacy in the age of medical big data. *Nature Medicine*, *25*(1), 37–43.

Ravindran, S., Suthar, J. K., Rokade, R., Deshpande, P., Singh, P., Pratinidhi, A., Khambadkhar, R., & Utekar, S. (2018). Pharmacokinetics, metabolism, distribution and permeability of nanomedicine. *Current Drug Metabolism*, *19*(4), 327–334.

Ristevski, B., & Chen, M. (2018). Big data analytics in medicine and healthcare. *Journal of Integrative Bioinformatics*, *15*(3), 20170030.

Sadoughi, F., Nasiri, S., & Ahmadi, H. (2018). The impact of health information exchange on healthcare quality and cost-effectiveness: A systematic literature review. *Computer Methods and Programs in Biomedicine*, *161*, 209–232.

Schachner, T., Keller, R., & Wangenheim, V. F. (2020). Artificial intelligence-based conversational agents for chronic conditions: Systematic literature review. *Journal of Medical Internet Research*, *22*(9), e20701.

Thakkar, V., & Gordon, K. (2019). Privacy and policy implications for big data and health information technology for patients: A historical and legal analysis. *Studies in Health Technology and Informatics*, *257*, 413–417.

Wang, F., & Preininger, A. (2019). AI in health: State of the art, challenges, and future directions. *Yearbook of Medical Informatics*, *28*(1), 16–26.

Xuan, J., Yu, Y., Qing, T., Guo, L., & Shi, L. (2013). Next-generation sequencing in the clinic: promises and challenges. *Cancer Letters*, *340*(2), 284–295.

Zeadally, S., Isaac, J. T., & Baig, Z. (2016). Security attacks and solutions in electronic health (E-health) systems. *Journal of Medical Systems*, *40*(12), 263.

Zhang, C., & Hua, Q. (2016). Applications of genome-scale metabolic models in biotechnology and systems medicine. *Frontiers in Physiology*, *6*, 413.

Zhang, J., Oh, Y. J., Lange, P., Yu, Z., & Fukuoka, Y. (2020). Artificial intelligence chatbot behavior change model for designing artificial intelligence chatbots to promote physical activity and a healthy diet: Viewpoint. *Journal of Medical Internet Research*, *22*(9), e22845.

Zhong, Y., Xu, F., Wu, J., Schubert, J., & Li, M. M. (2021). Application of next generation sequencing in laboratory medicine. *Annals of Laboratory Medicine*, *41*(1), 25–43.

4 Prediction of Pandemic and Epidemic Outbreak

Olugbemi T. Olaniyan
Rhema University, Aba, Nigeria

Charles Oluwaseun Adetunji
Edo State University, Uzairue, Nigeria

Olorunsola Adeyomoye
University of Medical Sciences, Ondo City, Nigeria

INTRODUCTION

Studies have indicated that the drivers for amplification and emergence of infectious diseases are anticipation, early detection, containment, and mitigation. Multidisciplinary responses are required to bring different solutions and emergency response to pandemic and epidemic infectious viruses (Pepin et al., 2010).

Modeling infectious diseases through artificial intelligence is beginning to gain tremendous interest among several biomedical scientists (Adetunji et al., 2020; Oyedara et al., 2022; Adetunji et al., 2022a, b, c, d, e, f, g, h; Olaniyan et al., 2022a, b). Over the past few decades modeling of different diseases like smallpox by Daniel Bernouli has been seen. The purpose of modeling is to understand the basic dynamics of an epidemic or outbreak so as to put in place control measures that will mitigate against the spread of the disease. Through accurate predictions, future outbreaks can inform preparedness, prevention approaches, and control strategies. Though there are some bottlenecks in infectious disease modeling such as mode of transmission form animals to humans, positive interactions with modelers are crucial in this regard.

The utilization of mathematical modeling is over a hundred years' old, but emerging diseases like Ebola, Zika virus, severe acute respiratory syndrome, multidrug-resistant malaria, West Nile virus, bioterrorism, and bird flu pandemic have attracted greater attention and importance to machine learning tools and algorithms. The primary aim of early warning signals is to facilitate the prediction of the magnitude and timing of an outbreak. It has been established during some studies that accurate forecasting could play a significant role during the time of an outbreak and pandemic or epidemic period (Damon, 2011). Meningitis, chicken pox, measles, diphtheria, poliomyelitis, mumps, rubella, smallpox, whooping cough, and scarlet fever have been predicted utilizing a mathematical model by Drake in 2006 through stochastic

DOI: 10.1201/9781003309468-4

modeling. In the study, it was discovered that final size of an outbreak is difficult to forecast due to disease specific situation and immediate local environment. The emergence of novel infectious disease agents mostly seems to be unpredictable. It is known that no single disease pathogen has been forecasted before the first discovery or appearance. The form of origin and the spread is the intrinsic part of the surveillance control measures. In recent years, the geographical location, wildlife, people, environment, and livestock interface are associated with the emergence of infectious diseases (Morse et al., 2012). The purpose of this chapter is to highlight various aspect of machine learning approach to predicting and forecasting epidemic and pandemics outbreaks.

PREDICTION OF PANDEMIC AND EPIDEMIC OUTBREAK

Morse et al. (2012) reported that prediction of an outbreak such as COVID-19 could be a daunting task which can be tackled by implementing simulation and modeling approaches. There are ecological theories that guide dynamic metapopulation problems such as Tipping points – Fisher information for analyzing critical transitions, approximating virus metapopulation dynamics coupled with Ma's time–diversity relationship adapted for the time–infection relationship and Hubbell's neutral theory, Taylor's power law, and Ma's population aggregation critical density for spatiotemporal stability/aggregation scaling. From their study, the authors reported that Taylor's power law and Ma's population aggregation critical density for spatiotemporal stability/aggregation scaling gave a better prediction model for many microbial infections giving accurate prediction value in terms of numbers and time. Cao et al. (2020) revealed that a dynamic model was established to predict infectious diseases particularly the time series model prediction of COVID-19. Using the mechanism of transmission of COVID-19, the control measures and prevention strategies were implemented and established for the dynamic time series model using various mathematical formulas. Analysis from their studies demonstrated that kinetic model and series analysis gave the summary of the cumulative diagnostic for COVID-19, regeneration rate, and infection rate. Based on the outcome of the results, the authors suggested that at the early stage of epidemic, emergency interventions like restriction, blocking, and supportive care will reduce the spread of the epidemic. Based on the epidemic, the authors generated data through collection to predict the epidemic and proffer necessary solution and preventive strategy. Cao et al. (2020) gave three approaches to predict the epidemic such as formulation of dynamic model, random process of statistical methods, and data mining techniques. Msmali et al. (2021) revealed that due to the wide spread and devastating effect of COVID-19 infection causing serious illness, artificial intelligence utilizing machine learning was implemented for the significant prediction of the COVID-19 pandemic. Using high-performance forecasting to accurately understand the dynamic of the disease and offer an accurate decision-making process through machine learning algorithms, certain recognition and prioritization were established. Machine learning algorithms like polynomial regression and support vector machine generated information such as recovered cases, number of confirmed cases and death cases. Based on artificial intelligence, machine learning applications are designed to provide automated programs, prediction, analysis, and data on detected pattern. Also, in a similar study conducted by Sujath et al. (2020), the authors revealed that machine learning algorithms can be utilized for the prediction of coronavirus disease. The authors demonstrated different machine learning models like multilayer perceptron, linear regression, and vector auto regression. In addition, Hamzah et al. (2020) reported that the COVID-19 outbreak can be predicted using data analysis and forecasting methods in a corona tracker which can increase response time to reduce the risk of COVID-19 like the SEIR model. Furthermore, Ardabili, et al. (2020) established that machine-learning models are very important algorithms for the prediction of pandemics like the COVID-19 outbreak. Through these models the authors carried out multilayer perceptron, fuzzy inference method, and adaptive network-based model demonstrating the highest simplification capability. Zhao et al. (2020) forecasted COVID-19 across different African countries utilizing the maximum-hunting parameter analysis approach and SEIR model to further enhance susceptible infected recovered. In other research carried out by Rustam et al. (2020), they proposed that increasing the decision-making ability in the face of COVID-19 pandemic, the machine-learning model should be utilized for the prediction and analysis of potential infected patients. In their study, machine learning like linear regression, vector support, exponential smoothing, and least absolute shrinkage were applied. From their results, it was suggested that exponential smoothing gave the highest prediction capacity of the pandemic. Xiang et al. (2021) reported that the COVID-19 pandemic spread across the globe could be predicted utilizing mathematical models. The authors performed public health intervention and prediction model of the pandemic and observed that most of the authors utilized the gamma distribution model, Erlang distribution, lognormal distribution and Weibull distribution. They were able to analyze the incubation period, infectious period, generation time, and serial interval and in the public health approaches, contact tracing, travel restriction, and social isolation had significant influence on prevention in addition to mask usage, and quarantine. Chen et al. (2021) analyzed the role of vaccine and antibody therapy in the treatment of COVID-19 and the role of mutation trackers in detecting over 5000 spike protein mutations. Utilizing deep learning, integrating genetics, biophysics and algebraic topology, the authors analyzed and identified 462 mutations on corresponding receptor-binding domains. Through their study, they were able to discover differences in mutation-induced binding free energy, weakened antibodies bindings and angiotensin-converting enzyme 2. Campillo-Funollet et al. (2021) reported that severe acute respiratory syndrome coronavirus 2 prediction was established utilizing an epidemiological model to forecast local outbreaks, guide healthcare demand, policy or decision-making process and public health decisions. The SEIR-D model utilized by the authors was able to generate counts of daily admissions, bed occupancy, and discharges from the first wave of COVID-19 infection through regional and local data. Prediction of the COVID-19 outbreak using machine learning algorithms is gaining more attention around the globe, particularly statistical models and epidemiological models, although there are several issues surrounding the effectiveness and efficiency in accurately predicting pandemic such as lack of standard models and essential data. It is known that accessibility to accurate prediction models for outbreaks is an essential such as soft computing models and machine learning like the adaptive network-based fuzzy inference system and multi-layered perceptron to generate data and information. Jonkmans et al. (2021) revealed that emerging diseases that can generate public health issues such as a pandemic which may not be sufficiently tackled by other preventive measures requires prompt attention and control through the utilization of available resources like pre-emptive prediction. The authors carried out a prediction survey by analyzing different emerging methods and discovered that stochastic models and spatiotemporal risk maps prediction are highly predominant with Rift Valley fever topping the chart of diseases. They revealed that diseases such as Ebola with complex socio-cultural risk factors are commonly predicted using multifactorial risk-based estimations. The authors demonstrated that there is a significant wide gap in forecasting and prediction models of future outbreaks of many priority diseases such as Ebola virus, Zika virus, Middle East respiratory syndrome, severe acute respiratory syndrome, Crimean-Congo hemorrhagic fever, Nipah and henipaviral disease, Lassa fever, and River Valley fever. It was discovered that few articles utilized spatio-temporal prediction but recommended room for improvement with numerous early warning signals by frontline epidemiological personnel and local health actors. Ahmed (2020) reported that different mathematical models are available to properly predict epidemiological and biological tendencies of pandemic or epidemic for prevention and control. The authors carried out a study to establish a mathematical modeling perspective in the COVID-19 spread and behavior change in Saudi Arabia. From their results, it was revealed that a weak positive relationship exists between mortality and spread of infection using susceptible-exposed-infection recovered model, which is a growth model focusing on infected, exposed, and recovered. They concluded from their study that social distancing, travel limitations, masking, and hygienic conditions are necessary to curtail the spread of the virus. Lone and Ahmad (2020) demonstrated that severe acute respiratory syndrome coronavirus-2 is high contagious, spreading rapidly across the globe with an increased number of deaths recorded daily. This resulted in a serious global pandemic requiring prompt measures and attention by all relevant stakeholders. They reported that Africa, referred to as the poorest continent with a large vulnerable population may be severely affected based on epidemiology, vulnerability, economic impact, preparedness, and etiological findings. Efforts in the last few decades towards large-scale epidemics have been channeled along the construction of several prediction models to help tackle the emerging diseases. There are different early warning models like Bayesian techniques which can combine data from diverse sources. Yadav and Akhter (2021) showed different prediction models and statistical techniques for the prediction of infectious diseases and outbreaks. The authors were able to demonstrate predictive monitoring approaches, time series modeling, epidemiological modeling, and distribution fitting. Yadav et al. (2020) revealed that data collected from infected individuals were utilized for prediction through prophet and daily average based algorithm for spread, death cases, infected ones, recovery, and future outbreaks. Singh et al. (2021) showed that the spread of COVID-19 across the globe is unprecedented, thus the prediction trend using public epidemiological data may be a preventive measure to the spread of this virus. The authors analyzed data from India, Pakistan, and Bangladesh using susceptible infection recovery, sequential Bayesian, exponential growth, time dependent, and maximum likelihood. From their study, it was demonstrated that the COVID-19 virus was spreading sub-continentally. Morse et al. (2012) reported that pandemics such as HIV/AIDS, influenza virus, and severe acute respiratory syndrome are known to drive ecological, socioeconomic, and behavioral changes. The authors described the prediction model based on statistical or mathematical modeling, communications, informatics, and diagnostic technologies for understanding the dynamics, trend, identify, pathogen discovery, and target surveillance. Ron and Chen (2021) reported that the global pandemic due to COVID-19 has ravaged many parts of the globe resulting in the death of millions. Through research, the authors were able to establish a prediction model by harnessing data and utilization of a machine learning approach for analysis. Regression models were employed to predict number of confirmed infected cases and spread were made to facilitate the decision-making process and possible optimal control measures. Castro et al. (2020) observed an unpredictable dynamic exponential growth pattern from infected individuals which might have been facilitated by epidemic spread of diseases. Susceptible infected removed models were utilized to analyze how lockdown measures influenced infection spread. Massaro et al. (2018) reported that managing large data derived from epidemics may be cumbersome, expensive, and very difficult. Thus, the theory of resilience was utilized to characterize the individual risk factor, system's functionality. This provided information about the impact of epidemics on population such as containment interventions. Xiang et al. (2021) noted that the developmental trend in the spread of the COVID-19 virus and public health intervention strategies can be analyzed through prediction using mathematical model. Giuliani et al. (2020) noted that severe acute respiratory syndrome coronavirus 2 has become a health emergency across the globe due to the severity of the virus. They revealed that the spatial areal aggregation model was utilized to analyze the spatiotemporal distribution, understand, and predict effectively spatiotemporal diffusion. They suggested that this model of prediction will assist the public decision-making process and also plan health policy programs. Pizzuti et al. (2020) reported the prediction accuracy using the susceptible–infectious–recovered epidemic model for SARS-CoV-2 virus across the region of Italy. From their study, it was revealed that a network-based model is a better prediction model for spread and identification of infected individuals with SARS-CoV-2 virus. Moein et al. (2021) showed a contradicting report on the use of a classical susceptible–infected–recovered model in the prediction of COVID-19. From their report, it was demonstrated that classical susceptible-infected-recovered model was unable to accurately predict or forecast the pattern and spread of epidemic due to inconformity and assumptions. Sinkala et al. (2021) reported that attenuating COVID-19 involve predicting the spread of the virus utilizing susceptible-infectious-recovered mathematical model. The authors revealed that mathematical models are an efficient approach such as a logistic model for predicting pandemics. Some reports have shown that the spread of COVID-19 virus through prediction using mathematical models is due to religious congregation, low testing rates, social contact structure, globalized trade and travel, climate change, agricultural practices, urbanization, and low identification of infected individuals. Zagrouba et al. (2020) analyzed the prediction of COVID-19 through the application of artificial intelligence-motivated methodologies. The authors utilized a predictive framework like support vector machines and neural network in forecasting COVID-19 outbreaks. Vytla et al. (2020) noted that COVID-19 pandemic prediction using a mathematical model such as Gaussian model, SEIRD, SEIHRD techniques, and SIRD influence the non-pharmaceutical intervention, risk factors understanding, and control of medical resources.

CONCLUSION

Global pandemic control and prevention approaches have evolved in the past few decades resulting in the utilization of more accurate and sophisticated methods like machine learning and computational tools. There is an advanced level of molecular approaches in pathogen discovery, identification, and surveillance through the internet, mobile phones, and digital devices. It has been established that adequate understanding of the process involved in the spreading of these diseases could form a basic prerequisite for adequate predictive analysis during a pandemic and epidemic period. Thus, capacity must be built and sustained in this direction to better mitigate against future pandemics and epidemics through the use of big data.

REFERENCES

Adetunji, Charles Oluwaseun, Egbuna, Chukwuebuka, Oladosun, Tolulope Olawumi, Akram, Muhammad, Michael, Olugbenga, Olisaka, Frances Ngozi, Ozolua, Phebean, Adetunji, Juliana Bunmi, Enoyoze, Goddidit Esiro, & Olaniyan, Olugbemi (2021). Efficacy of phytochemicals of medicinal plants for the treatment of human echinococcosis. In *Neglected tropical diseases and phytochemicals in drug discovery*. DOI: 10.1002/9781119617143.ch8

Adetunji, Charles Oluwaseun, Mitembo, William Peter, Egbuna, Chukwuebuka, Narasimha Rao, G.M. (2020). In silico modeling as a tool to predict and characterize plant toxicity. In Andrew G. Mtewa, Chukwuebuka Egbuna, & G.M. Narasimha Rao (Eds.), *Poisonous plants and phytochemicals in drug discovery*. Wiley Online Library DOI: 10.1002/9781119650034.ch14

Adetunji, Charles Oluwaseun, Nwankwo, Wilson, Olayinka, Akinola Samson, Olugbemi, Olaniyan Tope, Akram, Muhammad, Laila, Umme, Olugbenga, Michael Samuel, Oshinjo, Ayomide Michael, Adetunji, Juliana Bunmi, Okotie, Gloria E., & Esiobu, Nwadiuto (Diuto) (2022h). Machine learning and behaviour modification for COVID-19. In *Medical biotechnology, biopharmaceutics, forensic science and bioinformatics* (1st ed.). First Published 2022. Imprint CRC Press. Pages 17. eBook ISBN 9781003178903. DOI: 10.1201/9781003178903-17

Adetunji, Charles Oluwaseun, Nwankwo, Wilson, Olayinka, Akinola Samson, Olugbemi, Olaniyan Tope, Akram, Muhammad, Laila, Umme, Samuel, Michael Olugbenga, Oshinjo, Ayomide Michael, Adetunji, Juliana Bunmi, Okotie, Gloria E., & Esiobu, Nwadiuto (Diuto) (2022f). Computational intelligence techniques for combating COVID-19. In *Medical biotechnology, biopharmaceutics, forensic science and bioinformatics* (1st ed.). First Published 2022. Imprint CRC Press. Pages 12. eBook ISBN 9781003178903. DOI: 10.1201/9781003178903-16

Adetunji, Charles Oluwaseun, Olaniyan, Olugbemi Tope, Adeyomoye, Olorunsola, Dare, Ayobami, Adeniyi, Mayowa J, Alex, Enoch, Rebezov, Maksim, Garipova, Larisa, & Shariati, Mohammad Ali. (2022a). eHealth, mHealth, and telemedicine for COVID-19 pandemic. In S.K. Pani, S. Dash, W.P. dos Santos, S.A. Chan Bukhari, & F. Flammini (Eds.), *Assessing COVID-19 and other pandemics and epidemics using computational modelling and data analysis*. Springer. DOI: 10.1007/978-3-030-79753-9_10

Adetunji, Charles Oluwaseun, Olaniyan, Olugbemi Tope, Adeyomoye, Olorunsola, Dare, Ayobami, Adeniyi, Mayowa J., Alex, Enoch, Rebezov, Maksim, Isabekova, Olga, & Shariati, Mohammad Ali (2022c). Smart sensing for COVID-19 pandemic. In S.K. Pani, S. Dash, W.P. dos Santos, S.A. Chan Bukhari, & F. Flammini (Eds.), *Assessing COVID-19 and other pandemics and epidemics using computational modelling and data analysis*. Springer. DOI: 10.1007/978-3-030-79753-9_9

Adetunji, Charles Oluwaseun, Olaniyan, Olugbemi Tope, Adeyomoye, Olorunsola, Dare, Ayobami, Adeniyi, Mayowa J., Alex, Enoch, Rebezov, Maksim, Koriagina, Natalia, & Shariati, Mohammad Ali (2022e). Diverse techniques applied for effective diagnosis of COVID-19. In S.K. Pani, S. Dash, W.P. dos Santos, S.A. Chan Bukhari, & F. Flammini (Eds.), *Assessing COVID-19 and other pandemics and epidemics using computational modelling and data analysis*. Springer. DOI: 10.1007/978-3-030-79753-9_3

Adetunji, Charles Oluwaseun, Olaniyan, Olugbemi Tope, Adeyomoye, Olorunsola, Dare, Ayobami, Adeniyi, Mayowa J., Alex, Enoch, Rebezov, Maksim, Petukhova, Ekaterina, & Shariati, Mohammad Ali (2022b). Machine learning approaches for COVID-19 pandemic. In S.K. Pani, S. Dash, W.P. dos Santos, S.A. Chan Bukhari, & F. Flammini (Eds.), *Assessing COVID-19 and other pandemics and epidemics using computational modelling and data analysis*. Springer. DOI: 10.1007/978-3-030-79753-9_8

Adetunji, Charles Oluwaseun, Olaniyan, Olugbemi Tope, Adeyomoye, Olorunsola, Dare, Ayobami, Adeniyi, Mayowa J., Alex, Enoch, Rebezov, Maksim, Petukhova, Ekaterina, & Shariati, Mohammad Ali (2022d). Internet of Health Things (IoHT) for COVID-19. In S.K. Pani, S. Dash, W.P. dos Santos, S.A. Chan Bukhari, & F. Flammini (Eds.), *Assessing COVID-19 and other pandemics and epidemics using computational modelling and data analysis*. Springer. DOI: 10.1007/978-3-030-79753-9_5

Adetunji, Charles Oluwaseun, Olugbemi, Olaniyan Tope, Akram, Muhammad, Laila, Umme, Samuel, Michael Olugbenga, Oshinjo, Ayomide Michael, Adetunji, Juliana Bunmi, Okotie, Gloria E., Esiobu, Nwadiuto (Diuto), Oyedara, Omotayo Opemipo, & Adeyemi, Folasade Muibat (2022g). Application of computational and bioinformatics techniques in drug repurposing for effective development of potential drug candidate for the management of COVID-19. In *Medical biotechnology, biopharmaceutics, forensic science and bioinformatics* (1st ed.). First Published 2022, Imprint CRC Press. Pages 14. eBook ISBN 9781003178903. DOI: 10.1201/9781003178903-15

Adetunji, Charles Oluwaseun, & Oyeyemi, Oyetunde T. (2022). Antiprotozoal activity of some medicinal plants against entamoeba histolytica, the causative agent of amoebiasis. In *Medical biotechnology, biopharmaceutics, forensic science and bioinformatics* (1st ed.). First Published 2022. Imprint CRC Press. Pages 12. eBook ISBN 9781003178903. https://www.taylorfrancis.com/chapters/edit/10.1201/9781003178903-20/antiprotozoal-activity-medicinal-plants-entamoeba-histolytica-causative-agent-amoebiasis-charles-oluwaseun-adetunji-oyetunde-oyeyemi

Ahmed, Sirage Zeynu (2020). Analysis and forecasting the outbreak of Covid-19 in Ethiopia using machine learning. *European Journal of Computer Science and Information Technology*, 8(4), 1–13.

Akram, Muhammad, Adetunji, Charles Oluwaseun, Egbuna, Chukwuebuka, Jabeen, Shaista, Olaniyan, Olugbemi, Ezeofor, Nebechi Jane, Anani, Osikemekha Anthony, Laila, Umme, Găman, Mihnea-Alexandru, Patrick-Iwuanyanwu, Kingsley, Ifemeje, Jonathan Chinenye, Chikwendu, Chukwudi Jude, Michael, Olisah Chinedu, & Rudrapal, Mithun (2021b). Dengue fever. In *Neglected tropical diseases and phytochemicals in drug discovery*. DOI: 10.1002/9781119617143.ch17

Akram, Muhammad, Mohiuddin, Ejaz, Adetunji, Charles Oluwaseun, Oladosun, Tolulope Olawumi, Ozolua, Phebean, Olisaka, Frances Ngozi, Egbuna, Chukwuebuka, Michael, Olugbenga, Adetunji, Juliana Bunmi, Hameed, Leena, Awuchi, Chinaza Godswill, Patrick-Iwuanyanwu, Kingsley, & Olaniyan, Olugbemi (2021a). Chapter 7: Prospects of phytochemicals for the treatment of helminthiasis. In *Neglected tropical diseases and phytochemicals in drug discovery* (1st ed.). Wiley. DOI: 10.1002/978111961 7143.ch7

Ardabili, S. F., Mosavi, A., Ghamisi, P., Ferdinand, F., Varkonyi-Koczy, A. R., Reuter, U., & Atkinson, P. M. (2020). Covid-19 outbreak prediction with machine learning. Available at SSRN 3580188.

Campillo-Funollet, Eduard, Van Yperen, James, Allman, Phil, Bell, Michael, Beresford, Warren, Clay, Jacqueline, Dorey, Matthew, Evans, Graham, Gilchrist, Kate, Memon, Anjum, Pannu, Gurprit, Walkley, Ryan, Watson, Mark, & Madzvamuse, Anotida (2021). Predicting and forecasting the impact of local outbreaks of COVID-19: Use of SEIR-D quantitative epidemiological modelling for healthcare demand and capacity. *International Journal of Epidemiology*, 2021, 1103–1113. DOI: 10.1093/ije/dyab106

Castro, Mario, Ares, Saúl, Cuesta, José A., & Manrubia, Susanna (2020). The turning point and end of an expanding epidemic cannot be precisely forecast. *PNAS*, 117(42), 26190–26196.

Damon, I. (2011). Status of human monkeypox: Clinical disease, epidemiology and research. *Vaccine*; 29(suppl 4), D54–D59.

Drake, J.M. (2006). Limits to forecasting precision for outbreaks of directly transmitted diseases. DOI: 10.1371/journal.pmed.0030003

Giuliani, Diego, Dickson, Maria Michela, Espa, Giuseppe, & Santi, Flavio (2020). Modelling and predicting the spatio-temporal spread of COVID-19 in Italy. *BMC Infectious Diseases*, 20, 700. DOI: 10.1186/s12879-020-05415-7

Hamzah, F. B., Lau, C. H., Nazri, H., Ligot, D. V., Lee, G., Tan, C. L., & Ong, C. H. (2020). CoronaTracker: worldwide COVID-19 outbreak data analysis and prediction. *Bulletin of the World Health Organization*, 1, 32.

Chen, Jiahui, Gao, Kaifu, Wang, Rui, & Wei, Guo-Wei (2021). Prediction and mitigation of mutation threats to COVID-19 vaccines and antibody therapies. *Chemical Science*, 12, 6929. DOI: 10.1039/d1sc01203g

Cao, Jinming, Jiang, Xia, & Zhao, Bin (2020). Epidemic prediction of COVID-19. *Advances in Biotechnology & Microbiology*, 15(4), 555916. DOI: 10.19080/AIBM.2020.15.555916

Jonkmans, N., D'Acremont, V., & Flahault, A. (2021). Scoping future outbreaks: A scoping review on the outbreak prediction of the WHO Blueprint list of priority diseases. *BMJ Global Health*, 6, e006623. DOI: 10.1136/bmjgh-2021-006623

Lone, Shabir Ahmad, & Ahmad, Aijaz (2020). COVID-19 pandemic – An African perspective. *Emerging Microbes and Infections*, 9(1), 1300–1308. DOI: 10.1080/22221751.2020.1775132

Ma, Zhanshan (Sam) (2020). Predicting the outbreak risks and inflection points of COVID-19 pandemic with classic ecological theories. *Advancement of Science 7*, 2001530. DOI: 10.1002/advs.202001530

Massaro, Emanuele, Ganin, Alexander, Perra, Nicola, Linkov, Igor, & Vespignani, Alessandro (2018). Resilience management during large-scale epidemic outbreaks. *Scientific Reports*, 8, 1859. DOI: 10.1038/s41598-018-19706-2

Moein, Shiva, Nickaeen, Niloofar, Roointan, Amir, Borhani, Niloofar, Heidary, Zarifeh, Javanmard, Shaghayegh Haghjooy, Ghaisari, Jafar, & Gheisari, Yousof (2021). Inefficiency of SIR models in forecasting COVID-19 epidemic: A case study of Isfahan. *Scientific Reports*, 11, 4725. DOI: 10.1038/s41598-021-84055-6

Morse, Stephen S., Mazet, Jonna A. K., Mark Woolhouse, Parrish, Colin R., Carroll, Dennis, Karesh, William B., Carlos Zambrana-Torrelio, W Lipkin, Ian, & Daszak, Peter (2012). Prediction and prevention of the next pandemic zoonosis. *Lancet*, 380, 1956–1965.

Msmali, Ahmed, Zico, Mutum, Mechai, Idir, & Ahmadini, Abdullah (2021). Modeling and simulation: A study on predicting the outbreak of COVID-19 in Saudi Arabia. *Hindawi Discrete Dynamics in Nature and Society*, 2021, 5522928. DOI: 10.1155/2021/5522928

Olaniyan, Olugbemi T., Adetunji, Charles O., Adeniyi, Mayowa J., & Hefft, Daniel Ingo (2022a). Machine learning techniques for high-performance Computing for IoT applications in healthcare. In *Deep learning, machine learning and IoT in biomedical and health informatics* (1st ed.). First Published 2022. Imprint CRC Press. Pages 13. eBook ISBN 9780367548445. DOI: 10.1201/9780367548445-20

Olaniyan, Olugbemi T., Adetunji, Charles O., Adeniyi, Mayowa J., & Hefft, Daniel Ingo (2022b). Computational intelligence in IoT healthcare. In *Deep learning, machine learning and IoT in biomedical and health informatics* (1st ed.). First Published 2022. Imprint CRC Press. Pages 13. eBook ISBN 9780367548445. DOI: 10.1201/9780367548445-19

Oyedara, Omotayo Opemipo, Adeyemi, Folasade Muibat, Adetunji, Charles Oluwaseun, & Elufisan, Temidayo Oluyomi (2022). Repositioning antiviral drugs as a rapid and cost-effective approach to discover treatment against SARS-CoV-2 infection. In *Medical biotechnology, biopharmaceutics, forensic science and bioinformatics* (1st ed.). First Published2022.ImprintCRCPress.Pages12.eBookISBN9781003178903.DOI:10.1201/9781003178903-10

Pepin, K.M., Lass, S., Pulliam, J.R.C., Read, A.F., & Lloyd-Smith, J.O. (2010). Identifying genetic markers of adaptation for surveillance of viral host jumps. *Nature Reviews. Microbiology*, 8, 802–813.

Pizzuti, Clara, Socievole, Annalisa, Prasse, Bastian, & Van Mieghem, Piet (2020). Network-based prediction of COVID-19 epidemic spreading in Italy. *Applied Network Science*, 5, 91. DOI: 10.1007/s41109-020-00333-8

Rustam, F., Reshi, A. A., Mehmood, A., Ullah, S., On, B., Aslam, W., & Choi, G. S. (2020). COVID-19 future forecasting using supervised machine learning models. *IEEE Access*.

Singh, B.C., Alom, Z., Hu, H., Rahman, M.M., Baowaly, M.K., Aung, Z., Azim, M.A., & Moni, M.A. (2021). COVID-19 pandemic outbreak in the subcontinent: A data driven analysis. *Journal of Personalized Medicine* 11, 889. DOI: 10.3390/jpm11090889

Sinkala, Musalula, Nkhoma, Panji, Zulu, Mildred, Kafita, Doris, Tembo, Rabecca, & Daka, Victor (2021). The COVID-19 pandemic in Africa: Predictions using the SIR model. *Fortune Journal of Health Science*, 4(4), 491–499. DOI: 10.26502/fjhs.038

Sujath, R., Chatterjee, J. M., & Hassanien, A. E. (2020). A machine learning forecasting model for COVID-19 pandemic in India. *Stochastic Environmental Research and Risk Assessment*, 1.

Vytla, Vishnu, Ramakuri, Sravanth Kumar, Peddi, Anudeep, Kalyan Srinivas, K, & Nithish Ragav, N. (2020). Mathematical models for predicting Covid-19 pandemic: A review. *Journal of Physics: Conference Series*, 1797, 012009. DOI: 10.1088/1742-6596/1797/1/012009

Yue Xiang, Yonghong Jia, Linlin Chen, Lei Guo, Bizhen Shu, & Enshen Long (2021). COVID-19 epidemic prediction and the impact of public health interventions: A review of COVID-19 epidemic models. *Infectious Disease Modelling*, 6, 324e342.

Yadav, D., Maheshwari, H., & Chandra, U. (2020). Outbreak prediction of covid-19 in most susceptible countries. *Global Journal of Environmental Science and Management*, 6(SI), 11–20.

Yadav, S.K., & Akhter, Y. (2021). Statistical modeling for the prediction of infectious disease dissemination with special reference to COVID-19 spread. *Frontiers in Public Health*, 9, 645405. DOI: 10.3389/fpubh.2021.645405

Zagrouba, Rachid, Khan, Muhammad Adnan, Atta-ur-Rahman, Muhammad Aamer Saleem, Muhammad Faheem Mushtaq, Abdur Rehman, & Khan, Muhammad Farhan (2020). Modelling and simulation of COVID-19 outbreak prediction using supervised machine learning. *Computers, Materials & Continua*, 66(3), 2398–2407. DOI: 10.32604/cmc.2021.014042

Zhao, Z., Li, X., Liu, F., Zhu, G., Ma, C., & Wang, L. (2020). Prediction of the COVID-19 spread in African countries and implications for prevention and controls: A case study in South Africa, Egypt, Algeria, Nigeria, Senegal and Kenya. *Science of the Total Environment*, 729, 138959. DOI: 10.1016/j.scitotenv.2020.138959

5 Artificial Intelligence and Deep Learning in Treatment of Diseases Using Bioinformatics

Frank Abimbola Ogundolie
Baze University, Abuja, Nigeria
Federal University of Technology, Akure, Nigeria

Michael O. Okpara and Tolulope Peter Saliu
Federal University of Technology, Akure, Nigeria

Charles Oluwaseun Adetunji
Edo State University, Uzairue, Nigeria

Adedeji Olufemi Adetunji
North Carolina A&T State University, Greensboro, North Carolina

5.1 INTRODUCTION

Today, artificial intelligence (AI) has already been used in revolutionizing the world today with applications in diverse industries such as chatbots, driverless cars, automated customer support, virtual assistants automated workflow (Buhalis & Moldavska 2021), and risk assessment (Firouzi et al. 2021). With recent advancements reported in healthcare industries, the application of AI in this sector has resulted in improving the health status of patients through improved healthcare deliveries, especially in more efficient hospital managements, prognosis, vaccine development (Firouzi et al. 2021), virtual assistant nurses (McGrow 2019; Buchanan et al. 2020; Clancy 2020; Hornung & Smolnik 2021), patient monitoring (Nesarikar et al. 2020; Kantipudi et al. 2021), aiding in better diagnosis of disease condition using available data, better drug development (Freedman 2019; Jang 2019; Kaushik & Raj 2020; Firouzi et al. 2021; Gallego et al. 2021), improvement in decision-making (Hwang et al. 2019; Lysaght et al. 2019), by aiding in

DOI: 10.1201/9781003309468-5

prioritizing attending to sick patients in acute care medicine (Lynn 2019). These have also been reshaping precision medicine.

AI can also be used to predict the susceptibility of a group of people, race, or certain population to a particular disease using the genetic information obtained from them and also give an idea on their response to a type of treatment for these groups of people. Generally, the emergence of AI technology has greatly reduced the risk of prolonged, laborious, and false diagnoses that can likely or sometimes occur in our respective healthcare facilities. This has opened up the algorithm world to health.

The vast applications of AI using deep machine learning in healthcare have resulted in a much improved and better healthcare system today which is efficient and effective owing to their cost-effective and time saving early disease detection, prevention and management leading to amazing access to healthcare today for several patients. Robotic surgeries and procedures (Hashimoto et al. 2018; Han & Tian 2019; Beyaz 2020; Etienne et al. 2020; Chang et al. 2021; Collins et al. 2021) are now being engaged with minimized complications, and all these are as a result of the big data available to deep machine learning and utilized through deep machine learning algorithms.

With the advancements in telecommunication industries through the deployment of amazing technologies like the 4G and now 5G networks, the use of remote interactions in healthcare delivery through the use of various health tracking applications and software has further improved the health sector because consultation and on-demand healthcare services can be provided anytime and anywhere in the world today (Bohr & Memarzadeh 2020).

5.2 APPLICATION OF ARTIFICIAL INTELLIGENCE AND BIOINFORMATICS IN MEDICINE

The role AI and bioinformatics play in medicine in recent times cannot be overemphasized. While bioinformatics focuses on the management and analysis of biological data (Gusfield 2004), AI enables human-like reasoning (Korteling et al. 2021). Research in bioinformatics has been facilitated by the use of Information Technology in exploring genome-related aspects of biotechnology, from revolutionizing drug discovery processes, interpretation of medical images, decision support systems in diagnostics and therapy optimization, robotic surgery, repurposing of a drug, to designing novel vaccines (Jacobs et al. 2007; Levin et al., 2020). Overall, they have not only helped reduce the 14 years' drug development process to 4 or 5 years but also cut costs to about one-tenth. In addition, complex medical challenges such as the treatment of prenatal medical concerns at the genetic level are now possible (Debouck & Metcalf 2000).

5.3 ROLE OF AI IN DISEASE DIAGNOSIS

To ensure proper treatment and the well-being of patients, disease diagnosis has to be done properly. However, healthcare experts are burdened by the dynamic and changing environments of the medical system (Menschner et al. 2011). To mitigate these challenges and ensure that medical professionals keep up with constantly evolving diseases and changes in patient dynamics, AI has been deployed. The core role of AI during disease diagnosis is to find out if a patient is affected by a disease or not (Ransohof & Feinstein 1978).

To understand the vast extent to which AI simplifies the disease diagnosis process, it is vital to understand the procedure of disease diagnosis before AI intervention. When an individual suffers a health challenge, they reach out to a healthcare system where information regarding the patient is collected. Patient information collection is carried out by reviewing the patient's medical history, collecting vitals, and conducting physical exams after an interview. Thereafter, the patient is scheduled or referred for consultation. This highly complex process allows for errors due to individual objectiveness, experience, disposition or the mental state of the medical personnel (Chang & Hsu 2010). However, AI is designed to meet the dynamism of modern medical practice. One way by which it is supporting medical experts is in the early diagnosis of ectopic pregnancy and in providing a treatment option.

More so, a clinical support system based on AI algorithms is also used to assist a medical expert in selecting an accurate initial treatment to avoid future complications. Some of the AI algorithms used is the Support Vector Machine (SVM) and Multilayer Perceptron and they have helped gynecologists in making decisions about initial treatment with accuracy, sensitivity, and specificity of 96.1%, 96% and 98%, respectively (De Ramón Fernández et al. 2019). In addition, the use of supervised learning, a classification of the technology that learns associations based on existing sample or training data has been used in the diagnosis of dementia and cancer using the logistic regression model and voting strategy respectively (Rauschert et al. 2020).

5.4 ROLE OF AI IN PERSONALIZED MEDICINE

Due to genetic makeup and differential factors such as age and lifestyle, it is not a surprise to see a drug that works for one person be less effective on another or result in complications. To this end, medicine must be approached with the understanding that individuals are unique and treatment should be tailored to a person's medical history and biology. This is what led to the term personalized medicine (Ozomaro et al. 2013). To facilitate the personalized medicine approach, the genetic information of the individual is required. This serves as baseline data for tailoring medical treatment to a patient. It is to this end that the use of AI is important in ensuring precision, the accuracy of disease discovery, treatment, and drug administration.

Furthermore, computers are now being used in health facilities to record medical activities. In addition, the electronic health record (EHR) systems provide data that can be used to enhance medical service delivery. Other frequently used AI algorithms in personalized medicine include Naïve Bayesian, SVM, and the Artificial Neural Network (Ghumbre et al. 2011). To say the least, these algorithms have facilitated the achievement of accurate diagnosis, detection/prediction and treatment optimization. (Khan et al. 2013). Finally, successful implementation of current personalized medicine and future trends such as robotic surgeries would save lives and improve the medical profession.

5.5 USE OF PHARMACOGENOMICS FOR DRUG ADMINISTRATION

The human genome is made up of about 3.1 billion nucleotide base pairs and at least 22,000 genes which encode for different structural and functional proteins in humans (Collins et al. 2004; Piovesan et al. 2019; Salzberg 2018). The complete mapping of the human genome was completed in 2003 after 13 years of collaborative research involving research institutions, companies, and laboratories across Europe, Asia, and North America. The human genome project identified sites in the genome where genetic variations occur in humans and these genetic variations contribute to the differences in drug response seen in different individuals (Al-Ali et al. 2018). Consequently, many researchers have been motivated to carry out more studies in the fields of pharmacogenomics and personalized medicine.

Pharmacogenomics is a field in genomic medicine that studies how a patient's genetic composition affects their response to a therapeutic intervention. During the prescription and administration of drugs, epigenetic factors such as age, gender, weight, lifestyle, expression levels of certain liver and kidney enzymes, and family disease history are usually considered. For instance, the relatively higher rate of drug metabolism in young individuals compared to aged people could reduce the probability of experiencing adverse drug effects in young individuals. Also, the hormonal differences in male and female patients could cause a variation in drug response between male and female patients.

Though the patient's information informs the physician's drug prescription/administration, the genetic features of the patient contribute immensely to how the patient responds/reacts to a drug. This explains why different persons respond differently to the same dosage of a particular therapeutic intervention. The detection of genetic variation – even among people from the same race – and the role it plays in their different responses to the same drug have necessitated drug prescription and administration for an individual patient (personalized medicine).

Applying pharmacogenomics during drug prescription and administration has greatly enhanced the safety and efficacy of drugs while also reducing cases of drug toxicity, side effects, and reactions. Many studies geared towards identifying genetic variants associated with various diseases have been conducted (Bomba et al. 2017).

However, there is a need to conduct more studies and to make an inventory of disease-associated genetic variants responsible for variation in drug response in different persons. Genetic variations among people from different regions and among people from the same region affect the therapeutic efficacy of drugs administered to them. It is often assumed that racial differences are the chief determinant of genetic variation with people of different races showing differential expression of certain genes (especially genes associated with certain diseases). However, genetic variations among people from the same racial group are also not impossible. And if the dysregulated gene is critical for the manifestation of a particular disease, then it will not be unexpected to have differences in drug response among the patients.

In many cases, single nucleotide polymorphism(s) (SNPs) or single point mutation(s) in a gene associated with a disease could be responsible for variations in drug response in different patients. As it stands, researchers have identified over

600 million SNPs in human populations all over the globe (Floris et al. 2020). Another form of genetic variation is seen in insertions and deletions which can alter the genetic feature of an individual. These genetic variations could alter the pharmacodynamics and pharmacokinetics of the drug, thereby altering the drug dosing, drug response, or outrightly making the drug toxic. Consequently, personalized medicine for the diagnosis of disease and administration of drugs tailored specifically for an individual patient becomes very essential.

Personalized (also known as precision medicine) is a term used to describe disease diagnosis and treatment for an individual or a group of people based on the patient's genetic and epigenetic information. Advances in medicine have now made it possible to catalogue the genetic variations associated with disease and provide personalized medicine to patients based on their genetic features unlike in the past when the same treatment regimen for a particular disease is given to different people regardless of their genetic features. This has also substantially reduced the number of deaths associated with adverse effects of therapeutic drugs.

The pharmacokinetic properties of a drug is a major determinant of the efficacy and safety of the drug. The absorption, distribution, metabolism, elimination, and toxicity of a therapeutic drug in the patient are sometimes dependent on the physicochemical properties of the drug and the availability of efficient drug transporters and drug-metabolizing enzymes. Drug transporters are membrane-bound proteins that facilitate the influx of drugs into cells or efflux of drugs from cells. They are mostly expressed in the liver, kidney, brain, testis, and intestines. The role that drug transporters play in determining the pharmacokinetics of therapeutic drugs is enormous as the efficiency and toxicity of a drug can be determined by how quickly or slowly the drug is transported out of the cell. Many groups of drug transporters have been identified and well-studied. Most notable among them are the organic anion transporters (OAT 1, 2, and 3), organic cation transporters (OCT 1, 2, 3, N1, and N2), organic anion transporting peptides (OATP), oligopeptide transporters (PEPT 1 and 2), multidrug resistance-associated proteins (MRP 1, 2, and 3), and multidrug resistance protein 1 (MDR1) (Guéniche et al. 2020).

MDR1, also known as ATP-binding cassette subfamily B1 (ABCB1), is one of the most utilized drug transporters because of its broad substrate specificity. MDR1 is a major efflux transporter in many cells assisting with the efflux of therapeutic drugs and other metabolites from the cell. Some SNPs in the *ABCB1* gene encoding MDR1 protein have been reported with the 3435 C > T polymorphism in exon 26 contributing to the misfolding of MDR1 protein and affecting its primary function of drug elimination from the cell (Kimchi-Sarfaty et al. 2007). Genetic variation in the MDR1 and other drug transporters can affect the pharmacokinetics of a particular drug thus leading to different drug responses in a population.

The rate of drug metabolism is controlled by a group of drug-metabolizing enzymes which are also susceptible to genetic variations. And in many cases, the different responses to the same drug by different patients are attributed to genetic variations in the genes encoding the drug-metabolizing enzymes. The majority of the drug-metabolizing enzymes belong to a superfamily of enzymes called cytochrome P450 (CYP450) enzymes (Gardiner & Begg 2006). There are at least 30 identified CYP450 enzyme isoforms that are involved in Phase I or Phase II drug metabolism.

SNPs in CYP450 genes can create genetic variants that can alter the rate of drug metabolism giving rise to patients that are either ultrarapid, extensive, intermediate, or poor metabolizers of the same therapeutic drug (Nebert et al. 1999). Some well-studied CYP450 drug-metabolizing enzymes include CYP1A1, CYP1A2, CYP2A6, CYP2B6, CYP3A5, CYP3A7, CYP2D6, CYP2C19, CYP2C9, UGT1A1, UGT1A7, UGT2B7, GSTA5, GSTP1, GSTZ1, NAT2 (Sukasem & Medhasi 2018).

CYP2D6 is an important member of the CYP450 drug-metabolizing enzymes superfamily which is principally expressed in the liver. Although CYP2D6 is only approximately 3% of the entire CYP450 enzyme in the liver, it is involved in the metabolism of at least 25% of all clinically approved drugs (Wang et al. 2009). CYP2D6 is encoded by the CYP2D6 gene located on chromosome 22q13.1. According to Pharmacogene Variation Consortium (PharmVar), CYP2D6 has at least 149 identified polymorphic alleles (PharmVar 2022). Crossing over between CYP2D6 polymorphic alleles can potentially generate genetic variants and cause an alteration in the CYP2D6 phenotype expressed in different individuals. The major CYP2D6 variant alleles include CYP2D6*2, CYP2D6*4, CYP2D6*5, CYP2D6*10, CYP2D6*17 and CYP2D6*41 (Ingelman-Sundberg 2005). The gene expression levels of CYP2D6 or the presence of CYP2D6 genetic variants differ among individuals from a different population.

Thus, the rate of metabolism of CYP2D6 drugs will differ among the patients with differential expression levels of CYP2D6 and its genetic variants. Patients with a high expression level of CYP2D6 will rapidly metabolize and excrete CYP2D6 drugs leading to reduced toxicity of the drugs while patients with a low expression level of CYP2D6 will slowly metabolize CYP2D6 drugs thus making the drugs accumulate in the patient and increasing the risk of drug toxicity. CYP2D6 drug-metabolizing enzyme catalyzes the metabolism of a wide range of therapeutic agents including anti-depressants, antiarrhythmics, antiarrhythmicsphilic β-adrenoceptor blockers, and neuroleptics (Ingelman-Sundberg 2005). It is therefore important to evaluate the pharmacogenomics data of patients before prescribing/administering therapeutic drugs to improve drug efficacy and safety whilst also reducing adverse drug effects and toxicity.

5.6 AI IN MEDICINE – PAST AND PRESENT

The history of AI dates to 1950 when Alan Mathison Turing (the pioneer of machine learning) proved that computers can be programmed to think and process data like humans with what is now known as the "Turing Test." However, it was in the1970ss that AI was revolutionized to also find several applications in medicine. The incorporation of AI in medicine often involved interdisciplinary collaborations between experts in the fields of AI, computer science, and medicine. From the 1970s to this era, the application of AI in medicine has experienced steady and improved innovations from the use of AI to diagnose diseases based on stored information to the incorporation of AI in the development of medical robots for performing different kinds of surgeries.

The foundation for the application of AI in medicine can be traced back to 1971 when Prof Saul Amarel led a team of researchers at Rutgers University to set up the

Rutgers Research Resource on Computers in Biomedicine. The main objective for establishing the Rutgers Research Resource was to solve problems in biomedicine and psychology using AI while also building a strong network among biomedical scientists. Saul was the first principal investigator at the Rutgers Research Resource on Computers in Biomedicine; and during that time in 1974, Sholom Weiss became the first doctoral graduate in AI at Rutgers Research Resource. In 1978, Sholom Weiss, Kaz Kulikowski (an expert in medical modeling and decision-making) and Dr Aran Safir (an ophthalmologist) developed the Casual Associational Network (CASNET) model (Amarel 1985; Weiss et al. 1978). CASNET was designed to diagnose glaucoma and recommend treatments based on previous information about the disease for a particular patient. Building on the CASNET model, Weiss and Kulikowski collaborated to develop a more generalized system that can solve complex problems by imitating the decision-making capability of professionally trained human experts. This system was called the "EXPERT" and has been applied in many fields including Medicine (Weiss & Kulikowski 1979). The EXPERT system for application in medicine possesses a knowledge base that contains a wide range of disease features and patient information which the systems employ in giving diagnoses and recommending treatments for patients. The system's knowledge base can be regularly updated with new disease features and patient information by an expert in medicine to improve its overall performance.

The development of various AI systems with applications in medicine continued throughout the 1970s by researchers in different institutions. One of such AI systems is the INTERNIST-1 that was designed in 1972 by Dr Jack D. Meyer (a medical doctor), Dr Randolph A. Miller (a medical doctor) and Harry E. Pople, Jr (a computer scientist with a specialty in AI). The INTERNIST-1 (previously called DIALOG (Diagnostic logic)) system was designed to assist medical practitioners with the diagnosis of diseases which was a dilemma for practicing clinicians at the time. The diagnostic capability of the INTERNIST-1 system was built on intellectual diagnostic algorithms and a diversified diagnostic knowledge base which made it unique from other AI-assisted diagnostic tools (Miller et al. 1982). The INTERNIST-1 knowledge base contains the disease profile of various diseases, and this is the basis upon which the system diagnoses patients through different strategies like "diagnosis by exclusion."

A few years after the development of the INTERNIST-1 system, another AI system with application in medicine called MYCIN was developed. MYCIN was a "backward chaining" AI system developed in 1976 by Edward Hance Shortliffe (a computer scientist, biomedical informatician, and physician) at the Stanford University School of Medicine, California. The MYCIN system was designed to assist physicians with providing diagnosis and treatment suggestions for patients with bacterial infections. The application of MYCIN in bacterial infection diagnosis in the 1970s substantially reduced the wrong use of antibiotics by patients who had no access to experts in infectious diseases. The MYCIN system was operated on three sub-programs that work synergistically to provide a diagnosis for the patient, justify the suggested diagnosis and store the findings in the knowledge base of the system. The "Consultation sub-program" is the most important sub-program through which the physician supplies information about the patient to the system.

The "Explanation sub-program" gives justifications for the diagnosis provided based on the patient information supplied to the system. Then the "Knowledge acquisition sub-program" accepts and stores information supplied by experts in infectious diseases in the knowledge base of the MYCIN system (van Melle 1978). Over the years, the MYCIN system has been used to diagnose many other infectious diseases like meningitis and bacteremia.

The Present Illness Program (PIP) system was another AI applied in medicine for the diagnosis and treatment of oedema and other related diseases. It was developed in 1976 by Steven G. Pauker, Anthony G. Gorry, Jerome P. Kaissirer, and William B. Schwartz at the Massachusetts Institute of Technology. The organization or workflow of the PIP system is quite similar to that of MYCIN. The PIP system was built on four basic components.

(1) Specific clinical data about the patient that are supplied to the PIP system for processing.
(2) The supervisory program which collates the patient's clinical data, consults the short- and long-term memories of the system for past cases that are similar to that of the patient under investigation, then generates and tests hypotheses based on the information processed in the short- and long-term memories.
(3) The short-term memory is where the specific clinical data of the patient interacts with the medical knowledge uploaded into the long-term memory by a medical expert. The supervisory program ensures that only related and essential medical knowledge base from the long-term memory is processed for interaction with the specific clinical data from the patient under investigation.
(4) Long-term memory is where the knowledge base of the PIP system is contained and assembled into groups of related facts called "frames" depending on the characteristics of the disease (Pauker et al. 1976).

In 1986, a group of researchers in the computer science laboratory at the Massachusetts General Hospital developed a computer-based decision-support system capable of providing diagnostic assistance to physicians. Like the AI systems designed before it, DXplain accepts the patient's specific clinical data and searches through the system's enormous knowledge base for disease(s) that match the supplied clinical data. After some matches have been identified, the system suggests some diagnostic hypotheses and justifies its suggestions. The DXplain AI system has got a large knowledge base which has been upgraded from a knowledge base of about 500 diseases in 1986 to one that has descriptions for at least 2600 diseases and 5700 clinical findings (Barnett et al. 1987; Hoffer et al. 2005) (The Massachusetts General Hospital Computer Science Laboratory 2017). The DXplain AI system is user-friendly, possesses comprehensive medical information in its knowledge base, can justify its diagnostic hypotheses which are relatively more accurate, can be accessed remotely from different locations, and the knowledge base can easily be updated and improved (Barnett et al. 1987).

Advances and innovations in AI continued into the new millennium and beyond; especially as the knowledge base of most AI systems were updated constantly. In 2007, Watson®'s AI system was built by a team of researchers and engineers led by David Ferrucci at IBM (Ferrucci et al. 2013). IBM Watson® was originally designed to answer questions on a TV quiz show called *Jeopardy!* But with the steady improvement of the system, IBM Watson® has found applications in several fields in medicine and has been used for some novel findings in medicine. One of the fields in medicine where the IBM Watson® AI system has found application is in oncology. A double-blinded validation study was conducted by Somashekhar and his colleagues from 2014 to 2016 to ascertain the degree of concordance between treatment recommendations for breast cancer in 638 patients given by IBM Watson® and the treatment recommendations given by a multidisciplinary tumor board for breast cancer. They reported that 93% of the treatment recommendations given by IBM Watson® was consistent with the treatment recommendations given by the human experts (Somashekhar et al. 2018). This is an indication that the IBM Watson® AI system can be relied on for treatment suggestions. An upgrade of the knowledge base of the system could help broaden its application to other kinds of cancer.

In 2017, Nadine Bakkar and his colleagues at the Barrow Neurological Institute used IBM Watson® to screen all the RNA-binding proteins (RBPs) in the human genome and they identified six novel dysregulated RBPs associated with Lou Gehrig's disease (Bakkar et al. 2018). These RBPs could be potential biomarkers for the diagnosis of neurodegenerative disease. The study by Nadine Bakkar and his colleagues demonstrates that AI can be used as a veritable tool for novel discoveries in Medicine.

AI has also found application in the early detection and diagnosis of stroke-associated paralysis. In 2015, J. R. Villar in collaboration with S. Gonzalez, J. Sedano, C. Chira, and J. Trejo-Gabriel-Galan developed an AI-aided approach for early diagnosis of stroke. They used the "Genetic Fuzzy Finite State Machine" algorithm and a hybrid machine learning algorithm to detect stroke-associated paralysis. They also proposed that the AI system could be built into wearable devices for proper monitoring and early detection of stroke-associated paralysis (Villar et al. 2015). Today, wearable devices that could detect alterations in people's patterns of walking are now available, and they are very helpful for the early detection of stroke-associated paralysis (Bat-Erdene & Saver 2021; Mannini et al. 2016; Qiu et al. 2018).

AI has also been used to successfully restore motion in a quadriplegic patient. This was demonstrated in 2016 when Bouton and his colleagues used machine learning algorithms to understand the connection between the neuronal activity of the patient and the control activation of the muscles in his forearm through a neuromuscular electrical stimulation system. This system aided the quadriplegic patient to gain control of isolated motions of his fingers and hand (Bouton et al. 2016).

Researchers at Arterys have developed AI systems that are helpful to cardiologists, neurologists, and pulmonologists. CardioAI® is an AI system used in cardiology for monitoring the state of the heart (Arterys 2021a). Developed in 2017 by AI experts at Arterys, CardioAI® can produce very accurate, easy to interpret, and properly annotated rest, stress, and Holter electrocardiograms in a few seconds. The CardioAI® system has two main applications: the CardioAI® Application Program Interface which processes raw electrocardiograms and the CardioAI® Web platform or Graphic User Interface which displays the processed electrocardiograms for evaluation. With CardioAI®, remote cardiac monitoring of the patient is possible. Innovative improvements on the CardioAI® system have led to an extension of its application and the development of similar AI systems for other fields in Medicine. The Arterys LungAI® can be used to analyze chest computed tomography scans (Arterys 2021c). The Arterys Chest MSK AI® is an X-ray software that assists users to accurately read and interpret emergency radiographs in a short time (Arterys 2021b). Arterys NeuroAI® is a neuroradiology platform that assists neurologists with the diagnosis and treatment of brain tumors, stroke, multiple sclerosis, and some other neurodegenerative diseases (Arterys 2021d).

More recently in 2019, Roncato and his colleagues described an AI system that applies pharmacogenomics for precision medicine in Italy. FARMAPRICE was developed by two prominent IT companies with a mandate to computerize the Italian healthcare system. It is a pharmacogenomics clinical decision support system that was designed to ensure a more accurate prescription of efficacious drugs to patients through the collection of patients' genetic and epigenetic information. FARMAPRICE was integrated into the electronic health record system of Italy to promote the exchange of healthcare data between hospitals and to provide a much more accurate diagnosis and drug prescription based on the patient's genetic information. The FARMAPRICE AI system prototype is divided into four key sections for its application. The "Patients" section is a repository for genetic information of patients where new genetic data can be uploaded by the physician. The "Prescription" section suggests drug treatment for patients based on the genetic and epigenetic information provided. The "Drug configuration" and "File configuration" sections are reserved for periodic updating of the FARMAPRICE clinical decision support system with new pharmacogenomics guidelines (Roncato et al. 2019).

The application of AI in medicine has also been incorporated into medical surgeries with the advent of medical robots. Medical robots can now be used to perform surgeries in cardiology, neurology, urology, and other fields of medicine with high precision and accuracy. One such medical robots is the popular da Vinci® Surgical System, which was developed by scientists at Intuitive Surgical, Inc. The da Vinci® Surgical System received FDA approval in 2000 and has assisted surgeons to perform at least 6 million surgeries in cardiology, prostatectomy, nephrectomy, myomectomy, cystectomy, and hernia repair, etc. The da Vinci® Surgical System has a user-controlled console that enables the surgeon to control the robot, a patient cart that has four robotic arms for carrying out the procedures as controlled by the surgeon from the console and a 3D high-definition vision system which allows the operator to have a clear view of the surgical area (Azizian et al. 2011; Douissard et al. 2019).

Following the success of the da Vinci® Surgical System, similar AI-aided robotic systems for performing different surgical procedures have been developed by other IT companies. As essential as AI is, its application in medicine is to assist medical practitioners with providing more accurate medical diagnoses, better treatment options and improved patient management.

5.7 RISKS OF AI IN HEALTHCARE

Despite the numerous advantages of AI in the healthcare industry, there are still risks and threats these technologies possess to humans. This technology is largely dependent on big data. Availability of data can be a threat to this technology as many patients withhold information for privacy concerns and an attempt to enforce this might lead to several lawsuits from patients who believe their freedom of privacy has been infringed. This technology, unlike humans that are naturally prone to error of judgment, is also prone to error occasionally either through delayed servicing/ maintenance, technical problems, software problems, or problems due to data. These problems can result in wrong diagnosis and/or wrong management regime. If not detected early these can cause harm to patients.

Since AI depends on the available dataset, there has been an argument of the tendency for it to not recognize certain populations hence leading to bias in predicting, prevention, and management of such sets of people.

AI and deep machine learning have been able to transform the health sector with timely and precise disease diagnoses, effective management of diseases, and better treatment plans. This has made it possible to have treatments for precise or peculiar growth of community commonly referred to as precision medicine and have increased access to medics worldwide; the use of robotic surgeries and laparoscopy is now increasing the confidence of medical professionals worldwide. Despite these enormous advantages, these technologies can still be biased because it largely depends on data imputation and configuration.

REFERENCES

Al-Ali, M., Osman, W., Tay, G. K., & Alsafar, H. S. (2018). A 1000 Arab genome project to study the Emirati population. *Journal of Human Genetics*, 63(4), 533–536. https://doi.org/10.1038/s10038-017-0402-y

Amarel, S. (1985). Introduction to the comtex microfiche edition of the Rutgers University artificial intelligence research reports: The history Rutgers of artificial intelligence at Rutgers. *AI Magazine*, 6(3), 192–202.

Arterys. (2021a). Cardio AI. Retrieved February 3, 2022, from https://www.arterys.com/clinicalapp/cardioapp

Arterys. (2021b). Chest MSK AI. Retrieved February 3, 2022, from https://www.arterys.com/clinicalapp/chestmskapp

Arterys. (2021c). Lung AI. Retrieved February 3, 2022, from https://www.arterys.com/clinicalApp/lungapp

Arterys. (2021d). Neuro AI. Retrieved February 3, 2022, from https://www.arterys.com/clinicalapp/neuroapp

Azizian, M., Liu, M., Khalaji, I., & DiMaio, S. (2011). The da Vinci surgical system. In *Surgical robotics* (pp. 199–217). Springer. https://doi.org/10.1007/978-1-4419-1126-1_9

Bakkar, N., Kovalik, T., Lorenzini, I., Spangler, S., Lacoste, A., Sponaugle, K., … & Bowser, R. (2018). Artificial intelligence in neurodegenerative disease research: use of IBM Watson to identify additional RNA-binding proteins altered in amyotrophic lateral sclerosis. *Acta Neuropathologica*, 135(2), 227–247. https://doi.org/10.1007/s00401-017-1785-8

Barnett, G. O., Cimino, J. J., Hupp, J. A., & Hoffer, E. P. (1987). DXplain - An evolving diagnostic decision-support system. *JAMA*, 258(1), 67–74.

Bat-Erdene, B. O., & Saver, J. L. (2021). Automatic acute stroke symptom detection and emergency medical systems alerting by mobile health technologies: A review. *Journal of Stroke and Cerebrovascular Diseases*, 30(7), 105826. https://doi.org/10.1016/j.jstroke cerebrovasdis.2021.105826

Beyaz, S. (2020). A brief history of artificial intelligence and robotic surgery in orthopaedics & traumatology and future expectations. *Joint Diseases and Related Surgery*, 31(3), 653.

Bohr, A., & Memarzadeh, K. (2020). The rise of artificial intelligence in healthcare applications. *Artificial Intelligence in Healthcare*, 25–60. https://doi.org/10.1016/B978-0-12-818438-7.00002-2

Bomba, L., Walter, K., & Soranzo, N. (2017). The impact of rare and low-frequency genetic variants in common disease. *Genome Biology*, 18(77), 1–17. https://doi.org/10.1186/s13059-017-1212-4

Bouton, C. E., Shaikhouni, A., Annetta, N. V., Bockbrader, M. A., Friedenberg, D. A., Nielson, D. M., Sharma, G., Sederberg, P. B., Glenn, B. C., Mysiw, W. J., & Morgan, A. G. (2016). Restoring cortical control of functional movement in a human with quadriplegia. *Nature*, 533(7602), 247–250. https://doi.org/10.1038/nature17435

Buchanan, C., Howitt, M. L., Wilson, R., Booth, R. G., Risling, T., & Bamford, M. (2020). Predicted influences of artificial intelligence on the domains of nursing: Scoping review. *JMIR Nursing*, 3(1), e23939.

Buhalis, D., & Moldavska, I. (2021). In-room voice-based AI digital assistants transforming on-site hotel services and guests' experiences. In Wolfgang Wörndl, Chulmo Koo, Jason L. Stienmetz (Eds.) *Information and communication technologies in tourism 2021* (pp. 30–44). Springer.

Chan, W. K. V., & Hsu, C. (2010). How hyper-network analysis helps understand human networks?. *Service Science*, *2*(4), 270–280.

Chang, T. C., Seufert, C., Eminaga, O., Shkolyar, E., Hu, J. C., & Liao, J. C. (2021). Current trends in artificial intelligence application for endourology and robotic surgery. *Urologic Clinics*, 48(1), 151–160.

Clancy, T. R. (2020). Technology solutions for nurse leaders. *Nursing Administration Quarterly*, 44(4), 300–315.

Collins, F. S., Lander, E. S., Rogers, J., Waterston, R. H., & Conso, I. H. G. S. (2004). Finishing the euchromatic sequence of the human genome. *Nature*, 431(7011), 931–945. https://doi.org/10.1038/nature03001

Collins, J. W., Marcus, H. J., Ghazi, A., Sridhar, A., Hashimoto, D., Hager, G., Arezzo, A., Jannin, P., Maier-Hein, L., Marz, K., Valdastri, P., Mori, K., Elson, D. Giannarou, S., Slack, M., Hares, L., Beaulieu, Y., Levy, J. Laplante, G., Ramadorai, A., Jarc, A., Andrews, B., Garcia, P., Neemuchwala, H., Andrusaite, A., Kimpe, T., Hawkes, D., John, D., Kelly, J., & Stoyanov, D. (2021). Ethical implications of AI in robotic surgical training: A Delphi consensus statement. *European Urology Focus*, 8(2), 613–622. https://doi.org/10.1016/j.euf.2021.04.006

Debouck, C., & Metcalf, B. (2000). The impact of genomics on drug discovery. *Annual Review of Pharmacology and Toxicology*, *40*, 193–207. https://doi.org/10.1146/annurev.pharmtox.40.1.193

Douissard, J., Hagen, M. E., & Morel, P. (2019). The da Vinci surgical system. In Jonathan Douissard, Monika E. Hagen & P. Morel (Eds.) *Bariatric robotic surgery* (pp. 13–27). Springer, Cham. https://doi.org/10.1007/978-3-030-17223-7_3

Etienne, H., Hamdi, S., Le Roux, M., Camuset, J., Khalife-Hocquemiller, T., Giol, M., Debrosse, D. and Assouad, J. (2020). Artificial intelligence in thoracic surgery: past, present, perspective and limits. *European Respiratory Review*, 29(157).

Fernández, A. D. R., Fernández, D. R., & Sanchez, M. T. P. (2019). A decision support system for predicting the treatment of ectopic pregnancies. *International journal of medical informatics*, *129*, 198–204.

Ferrucci, D., Levas, A., Bagchi, S., Gondek, D., & Mueller, E. T. (2013). Watson: Beyond Jeopardy! *Artificial Intelligence*, 199–200, 93–105. https://doi.org/10.1016/j.artint.2012.06.009

Firouzi, F., Farahani, B., Daneshmand, M., Grise, K., Song, J., Saracco, R., Wang, L.L., Lo, K., Angelov, P., Soares, E., & Luo, A. (2021). Harnessing the power of smart and connected health to tackle covid-19: IoT, AI, robotics, and blockchain for a better world. *IEEE Internet of Things Journal*, 8(16), 12826–12846.

Floris, M., Cano, A., Porru, L., Addis, R., Cambedda, A., Idda, M. L., … Maioli, M. (2020). Direct-to-consumer nutrigenetics testing: An overview. *Nutrients*, 12(2), 1–13. https://doi.org/10.3390/nu12020566

Freedman, D. H. (2019). Hunting for new drugs with AI. *Nature*, 576(7787), S49–S53.

Gallego, V., Naveiro, R., Roca, C., Ríos Insua, D., & Campillo, N. E. (2021). AI in drug development: A multidisciplinary perspective. *Molecular Diversity*, 25(3), 1461–1479.

Gardiner, S. J., & Begg, E. J. (2006). Pharmacogenetics, drug-metabolizing enzymes, and clinical practice. *Pharmacological Reviews*, 58(3), 521–590. https://doi.org/10.1124/pr.58.3.6.521

Ghumbre, S., Patil, C., & Ghatol, A. (2011, December). Heart disease diagnosis using support vector machine. In *International conference on computer science and information technology (ICCSIT')* 17–18 December 2011; Pattaya, Thailand, pp. 84–88.

Guéniche, N., Bruyere, A., Le Vée, M., & Fardel, O. (2020). Implication of human drug transporters to toxicokinetics and toxicity of pesticides. *Pest Management Science*, 76(1), 18–25. https://doi.org/10.1002/ps.5577

Gusfield, D. (2004). Introduction to the IEEE/ACM transactions on computational biology and bioinformatics. *IEEE/ACM Transactions on Computational Biology and Bioinformatics*, *1*(01), 2–3.

Han, X. G., & Tian, W. (2019). Artificial intelligence in orthopaedic surgery: current state and future perspective. *Chinese Medical Journal*, 132(21), 2521–2523.

Hashimoto, D. A., Rosman, G., Rus, D., & Meireles, O. R. (2018). Artificial intelligence in surgery: promises and perils. *Annals of Surgery*, 268(1), 70.

Hoffer, E. P., Feldman, M. J., Kim, R. J., Famiglietti, K. T., & Barnett, G. O. (2005). DXplain: Patterns of use of a mature expert system. In *AMIA 2005 Annual Symposium Proceedings/AMIA Symposium* October 22, 2005 - October 26, 2005Hilton Washington & TowersWashington, DC (pp. 321–325).

Hornung, O., & Smolnik, S. (2021). AI invading the workplace: negative emotions towards the organizational use of personal virtual assistants. *Electronic Markets*, 1–16

Hwang, D. K., Hsu, C. C., Chang, K. J., Chao, D., Sun, C. H., Jheng, Y. C., Yarmishyn, A. A., Wu, J. C., Tsai, C. Y., Wang, M. L., Peng, C. H., Chien, K. H., Kao, C. L., Lin, T. C., Woung, L. C., Chen, S. J., & Chiou, S. H. (2019). Artificial intelligence-based decision-making for age-related macular degeneration. *Theranostics*, 9(1), 232–245. https://doi.org/10.7150/thno.28447

Ingelman-Sundberg, M. (2005). Genetic polymorphisms of cytochrome P450 2D6 (CYP2D6): Clinical consequences, evolutionary aspects and functional diversity. *Pharmacogenomics Journal*, 5(1), 6–13. https://doi.org/10.1038/sj.tpj.6500285

Jacobs, G. H., Williams, G. A., Cahill, H., & Nathans, J. (2007). Emergence of novel color vision in mice engineered to express a human cone photopigment. *Science*, *315*(5819), 1723–1725.

Jang, I. J. (2019). Artificial intelligence in drug development: clinical pharmacologist perspective. *Translational and Clinical Pharmacology*, 27(3), 87–88.

Kantipudi, M. V. V., Moses, C. J., Aluvalu, R., & Kumar, S. (2021). Remote Patient Monitoring Using IoT, Cloud Computing and AI. In Akash Kumar Bhoi, Pradeep Kumar Mallick, Mihir Narayana Mohanty, and Victor Hugo C. de Albuquerque (Eds.) *Hybrid Artificial Intelligence and IoT in Healthcare* (pp. 51–74). Springer, Singapore.

Kaushik, A. C., & Raj, U. (2020). Ai-driven drug discovery: A boon against covid-19? *AI Open*, 1, 1–4.

Khan, A. N., Kiah, M. M., Khan, S. U., & Madani, S. A. (2013). Towards secure mobile cloud computing: A survey. *Future generation computer systems*, *29*(5), 1278–1299.

Kimchi-Sarfaty, C., Oh, J. M., Kim, I.-W., Sauna, Z., Calcagno, A. M., Ambudkar, S., & Gottesman, M. (2007). A "Silent" polymorphism in the MDR1 gene changes substrate specificity. *Science*, 315(November), 525–529. https://doi.org/10.1126/science.1135308

Korteling, J. H., van de Boer-Visschedijk, G. C., Blankendaal, R. A., Boonekamp, R. C., & Eikelboom, A. R. (2021). Human-versus artificial intelligence. *Frontiers in artificial intelligence*, *4*, 622364.

Levin, A. T., Hanage, W. P., Owusu-Boaitey, N., Cochran, K. B., Walsh, S. P., & Meyerowitz-Katz, G. (2020). Assessing the age specificity of infection fatality rates for COVID-19: systematic review, meta-analysis, and public policy implications. *European journal of epidemiology*, *35*(12), 1123–1138. https://doi.org/10.1007/s10654-020-00698-1

Lysaght, T., Lim, H. Y., Xafis, V., & Ngiam, K. Y. (2019). AI-assisted decision-making in healthcare. *Asian Bioethics Review*, 11(3), 299–314.

Lynn, L. A. (2019). Artificial intelligence systems for complex decision-making in acute care medicine: A review. *Patient Safety in Surgery*, 13(1), 1–8.

Mannini, A., Trojaniello, D., Cereatti, A., & Sabatini, A. M. (2016). A machine learning framework for gait classification using inertial sensors: Application to elderly, post-stroke and Huntington's disease patients. *Sensors (Switzerland)*, 16(1). https://doi.org/10.3390/s16010134

McGrow, K. (2019). Artificial intelligence: Essentials for nursing. *Nursing*, 49(9), 46.

Menschner, P., Prinz, A., Koene, P., Köbler, F., Altmann, M., Krcmar, H., & Leimeister, J. M. (2011). Reaching into patients' homes–participatory designed AAL services: The case of a patient-centered nutrition tracking service. *Electronic Markets*, *21*(1), 63–76.

Miller, R. A., Pople Jr, H. E., & Myers, J. D. (1982). INTERNIST-1, an experimental computer-based diagnostic consultant for general internal medicine. *The New England Journal of Medicine*, 307(8), 468–476.

Nebert, D. W., Ingelman-Sundberg, M., & Daly, A. K. (1999). Genetic epidemiology of environmental toxicity and cancer susceptibility: Human allelic polymorphisms in drug-metabolizing enzyme genes, their functional importance, and nomenclature issues. *Drug Metabolism Reviews*, 31(2), 467–487. https://doi.org/10.1081/DMR-100101931

Nesarikar, A., Haque, W., Vuppala, S., & Nesarikar, A. (2020). COVID-19 remote patient monitoring: Social impact of AI. *arXiv preprint arXiv:2007.12312.*

Pauker, S. G., Gorry, G. A., Kassirer, J. P., & Schwartz, W. B. (1976). Towards the simulation of clinical cognition. Taking a present illness by computer. *The American Journal of Medicine*, 60(7), 981–996. https://doi.org/10.1016/0002-9343(76)90570-2

Ozomaro, U., Wahlestedt, C., & Nemeroff, C. B. (2013). Personalized medicine in psychiatry: problems and promises. *BMC medicine*, *11*, 1–35.

PharmVar. (2022). CYP2D6. Retrieved February 3, 2022, from https://www.pharmvar.org/gene/CYP2D6

Piovesan, A., Pelleri, M. C., Antonaros, F., Strippoli, P., Caracausi, M., & Vitale, L. (2019). On the length, weight and GC content of the human genome. *BMC Research Notes*, 12(1), 1–7. https://doi.org/10.1186/s13104-019-4137-z

Qiu, S., Liu, L., Zhao, H., Wang, Z., & Jiang, Y. (2018). MEMS inertial sensors based gait analysis for rehabilitation assessment via multi-sensor fusion. *Micromachines*, 9(442), 1–17. https://doi.org/10.3390/mi9090442

Ransohoff, D. F., & Feinstein, A. R. (1978). Problems of spectrum and bias in evaluating the efficacy of diagnostic tests. *The New England journal of medicine*, *299*(17), 926–930.

Rauschert, S., Melton, P.E., Heiskala, A., Karhunen, V., Burdge, G., Craig, J.M., Godfrey, K.M., Lillycrop, K., Mori, T.A., Beilin, L.J. and Oddy, W.H. (2020). Machine learning-based DNA methylation score for fetal exposure to maternal smoking: development and validation in samples collected from adolescents and adults. *Environmental Health Perspectives*, *128*(9), 097003.

Roncato, R., Cin, L. D., Mezzalira, S., Comello, F., De Mattia, E., Bignucolo, A., … & Cecchin, E. (2019). FARMAPRICE: A pharmacogenetic clinical decision support system for precise and cost-effective therapy. *Genes*, 10(4). https://doi.org/10.3390/genes10040276

Salzberg, S. L. (2018). Open questions: How many genes do we have? *BMC Biology*, 16(94), 10–12. https://doi.org/10.1186/s12915-018-0564-x

Somashekhar, S. P., Sepúlveda, M. J., Puglielli, S., Norden, A. D., Shortliffe, E. H., Rohit Kumar, C., … Ramya, Y. (2018). Watson for oncology and breast cancer treatment recommendations: Agreement with an expert multidisciplinary tumour board. *Annals of Oncology*, 29(2), 418–423. https://doi.org/10.1093/annonc/mdx781

Sukasem, C., & Medhasi, S. (2018). Clinical pharmacogenomics and personalized medicine: New strategies to maximize drug efficacy and avoid adverse drug reaction. In Y. Pathak (Ed.), *Genomics-driven healthcare: Trends in disease prevention and treatment* (pp. 239–261). Adis, Singapore. https://doi.org/10.1007/978-981-10-7506-3_13

The Massachusetts General Hospital Computer Science Laboratory. (2017). DXplain®. Retrieved February 2, 2022, from http://www.mghlcs.org/projects/dxplain

van Melle, W. (1978). MYCIN: A knowledge-based consultation program for infectious disease diagnosis. *International Journal of Man-Machine Studies*, 10(3), 313–322. https://doi.org/10.1016/S0020-7373(78)80049-2

Villar, J. R., González, S., Sedano, J., Chira, C., & Trejo-Gabriel-Galan, J. M. (2015). Improving human activity recognition and its application in early stroke diagnosis. *International Journal of Neural Systems*, 25(4), 1–20. https://doi.org/10.1142/S0129065714500361

Wang, B., Yang, L.-P., Zhang, X.-Z., Huang, S.-Q., Bartlam, M., & Zhou, S.-F. (2009). New insights into the structural features and functional relevance of human cytochrome P450 2C9 enzyme. *Drug Metabolism Reviews*, 41(4), 573–643. https://doi.org/10.3109/03602530903118729

Weiss, S. M., & Kulikowski, C. A. (1979). EXPERT: A system for developing consultation models. In *6th International Joint Conference on Artificial Intelligence, Tokyo Japan August 20–23, 1979* (pp. 942–947).

Weiss, S. M., Kulikowski, C. A., & Amarel, S. (1978). A model-based method for computer-aided medical decision-making. *Artificial Intelligence*, 11, 145–172.

6 Personalized and Public Health Systems

Hannah Edim Etta
University of Cross River State, Calabar, Nigeria

6.1 INTRODUCTION

6.1.1 Personalized Health Systems

There are several definitions of personalized health systems (PHS) but in general PHS are designed to provide individual tailored healthcare and, at the same time, enable the public healthcare system to offer good quality care to large populations and still maintain a self-sustaining system. The self-sustaining system includes several services used to solve the needs of each and every individual patient. Hence the promotion of the Mobile Health tool (mhealth) as a reliable tool for health care and delivery of public health services (WHO, 2016). In recent times too, there's a call for countries to develop and use digital technology for advancing universal health coverage. This aspect of medicine is still very new, especially in countries like Nigeria. Apart from the fact that it may not even be practiced in many localities, is the fact that the majority of Nigerians are not IT literate enough to handle and operate simple digital health devices on their own, and those that are knowledgeable may not be able to afford it.

PHS, also known as precision medicine, makes it possible for healthcare providers to deliver targeted treatment and prevention plans to patients. It involves incorporating the unique genetic makeup of an individual, their environment, and life style choices into diagnosing and managing their health challenges.

6.2 PERSONALIZED HEALTHCARE TOOLS

With the myriads of digital tools like big data analytics, kits for genetic testing, electronic health records, and supercomputing, physicians, other medical practitioners and scientists are breaking frontiers in personalized medicine. Many countries have set up precision medicine initiatives for their citizens (about 40 countries) with China's recorded as being the largest (WEF, 2022).

At the onset of digitalization, electronic health records (EHR) were used for PHS. These were used to develop decision support system (DSS) with complete guidelines for users, online. Currently, real-time data is used to code predictive models, for identifying patients at risk, and for providing personalized care and prompt intervention too. EHR and management systems are employed in rendering these services.

 DOI: 10.1201/9781003309468-6

This has greatly improved clinical outcomes and has been seen to be cost-effective (Friedman et al., 2003). Telemedicine is another personalized healthcare tool that has been used in recent times, where telecommunications and electronic means are employed to distribute health-related services and information to patients and the general public. This makes it easier for the healthcare provider to reach the patient across distances for care, advice, education, intervention, reminders, follow-up checkups and remote admission as the case may be.

In recent times, however, PHS tools have become more digital with several online medical services like pediatrics online, general practitioners online, gynecologists online, etc. There's also teledermatology, a completely novel tool on its own. These tools bring about immediate improved health services and high quality clinical care. Analytical tools using big data obtained from patients based on their behavior, emotions, genomics patient-reported outcome (PRO), combined with clinical (lab, medications, and diagnosis) data, are being researched presently (Lewy et al., 2019). With the sequencing of the human genome, genomic data of individual patients is at the fingertips of researchers and is a relatively new addition to the PHS clinician's tool box (Bresnick, 2018).

Individuals (patients) themselves, also take part in the design of new services and how they are delivered, and how the outcomes are measured. In Canada and other such climes today, you can call up an agent to link you up with a family doctor, gynecologist, surgeon, or even a vet. Health maintenance organizations (HMOs) have been set up as commercial entities to collaborate with industries and the academia to design the implementation and evaluation of PHS. This takes care of the problem of shortages of economic and human resources. Some of the interventions by HMOs are to implement models that are predictive and to identify patients at high risk (HR) and focus on possible cures. These models have been proven to be applicable and highly effective over time with large populations. A typical example is reported by Goshen et al., 2017. In 2012, a proactive personalized care plan for HR renal failure based on predictive modelling for identifying individuals at risk, was engaged and improved outcomes and cost-effectiveness were reported. Predictive modelling was also implemented for colon cancer in 2013, based on routine blood tests that identifies individuals that will benefit from colorectal screening for cancer routinely.

In 2015, President Barrack Obama, former President of the United States of America, launched an initiative known as the Precision Medicine Initiative (PMI). PMI collects health data from Americans and saves them in a biobank. These data are used for designing personalized or precision treatments for individuals. There are many other such projects springing up in developed climes. Vigorous research among the academia has been ongoing to combine data science and breakthroughs in bioengineering with laboratory research and clinical discovery for producing precision medicine tools. There are also collaborations between pharmaceutical companies, research institutions and healthcare providers to leverage on these breakthroughs and tip in heavy investments for expanding their drug database.

As reported by Bresnick (2018), investors like Mark Zuckerberg and Bill Gates are also promoting strategies that promote personalized medicine to be implemented. Recently, Bill Gates announced a $50 million investment for data-driven research on Alzheimer's disease. Zuckerberg and his wife also recently donated $10 million to UCSF Institute for scientists to explore how best to use data that are already available to produce novel information. This same trend is being observed worldwide among developed economies.

As far back as 2015, the Chinese already commenced the use of precision medicine by developing computational power and artificial intelligence tools for discovering new drugs and treatment interventions and how to deliver the same to particular patients. For example, iCarbonX pharmaceutical, a Chinese company founded in 2015, collects health data on the environment, genetics and behaviors of millions of patients and uses the same to diagnose and formulate the best treatment and mode of delivery to individuals.

In 2016 also, Wuxi Nextcode and Huawei launched a new partnership developing a cloud computing infrastructure that is required to store massive amounts of data needed for personalized healthcare. With partnerships and companies like these, China is advancing precision medicine using fourth industrial revolution technologies (WEF, 2022).

All over Europe, there's a lot of pushing for investments, innovations, and research into precision medicine as advances in genomics have made it possible to diagnose and treat diseases like diabetes, cancer, and rare metabolic diseases. These genetic testing and biogenic tools have actually been in existence and well established in Europe, but it's only recently that they are being engaged for advanced diagnosis and therapeutics. European scientists have created a method of producing customized proteins using DNA sequences of individuals. These methods have been successfully applied for the development of therapies for insulin treatment and cancer treatment (MT, 2022).

In most African countries, the only tools available are hospital visit records. In Rwanda, the only available tool for accessing the pneumococcal conjugate vaccine (PCV) impact on pneumonia and pediatric meningitis hospitalizations (Gatera et al., 2016) were hand-recorded administrative records. For a typical resource-poor African health system, admission log books and data from routine surveillance will suffice for sophisticated impact studies.

In recent times, however, Nigeria is opening up to laboratories and bio-companies concerned with precision medicine. An ultra-modern genetic testing and preventive healthcare and wellness center was recently opened by SYNLAB Nigeria (www.vanguardngr.com/). A World Bank-funded research organization, The African Centre of Excellence for Genomics of Infectious Diseases (ACEGID), and another genomic outfit, 54Gene, both based in Nigeria, aim to produce genetic data of humans of African origin to gather enough samples of Africans so that they can work with individual researchers and pharmaceutical companies to develop drugs and treatments tailored for Africans, to treat diseases peculiar to Africans like sickle cell, diabetes, hypertension, etc. bearing in mind that the future is precision medicine (Edward-Ekpu, 2014).

Another PHS tool gaining wide acceptance in Nigeria is telemedicine. As stated earlier, telemedicine is related to using digital devices, mobile phones or computers to access healthcare. There are websites where patients can readily speak with a doctor via voice or video call. The COVID-19 pandemic in our recent history has made telemedicine popular and widely accepted in Nigeria since one can receive healthcare without necessarily going to the hospital and risk contracting an infection. It is, however, yet to get to the grassroots health system.

Machine learning is another tool that will open up healthcare delivery in Africa. Basically, with machine learning, a computer program is built to perform a task without giving it any instructions. The program uses complex data to identify patterns and from these make predictions, timely predictions about the named disease and clinical outcomes. These huge data (97%) have been lying around for decades in Nigeria untapped. It is hoped that machine learning in the near future will revolutionize healthcare systems and deliver proactive, predictive, easily available, and easily affordable personalized healthcare.

6.3 PROBLEMS HINDERING PERSONALIZED HEALTHCARE SYSTEMS IN AFRICA

Personalized Healthcare Systems face the following limitations in Africa:

i. Poverty: This is a major problem that may stall personalized medicine in Africa.
ii. Acceptability of personalized medicine by the majority of Africans is doubtful. Aside the problem of illiteracy, most Africans believe in reactive medicine rather than the preventive model.
iii. Again, like earlier stated, African countries lack the facilities and expertise needed to successfully implement personalized medicine. Before any kind of personalized healthcare can be made available to the African public/citizenry, there must be intensive training of doctors, nurses, and other health workers on the tools and technologies of how to effectively harness and understand the science and complex data involved in personalized medicine. These tools will be needed to communicate complex ideas to patients. Sensitive personal data of patients will have to be protected over the internet. Information about their genetics and biology will have to be communicated to them and their permission sought before such can be used.
iv. All stakeholders scientists, health workers, public/private partners, civil society, industry and policymakers, etc. will have to come together to untangle and broaden our understanding of how our body works, and how our genetics and biology affects disease.

Personalized medicine is expensive in all its ramifications and African countries may just not be ready for it yet.

6.4 PUBLIC HEALTH SYSTEMS

Public health as a science deals with improving, protecting and securing individuals', families', communities' and populations' health status. This intervention could be in small communities or in wide regions of the world.

Any institutions, public, private, or voluntary, that contribute to delivering essential public health services to the citizenry, within a jurisdiction, is a referred to as a public health system. They are very vital to any health systems.

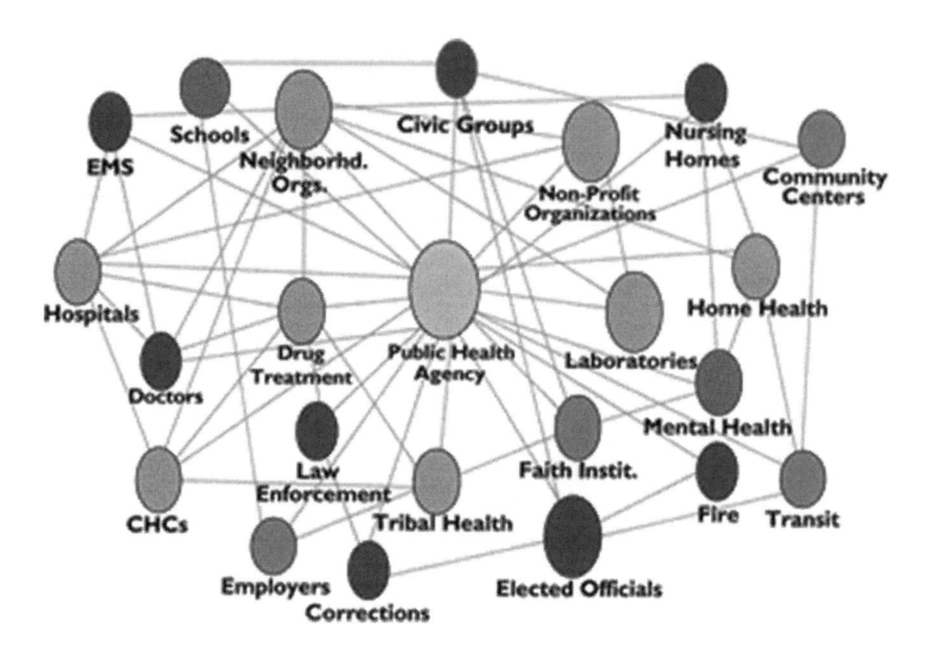

FIGURE 6.1 Public health systems.

Source: WWW.CDC.GOV

Generally, public health is about engaging a population and not just individuals in health activities. It has to do with undertaking a population's health approach based on the fact that genetic, behavioral, and socio-economic factors like education, financial status, housing, etc. influence the overall health and well-being of the population (Jarvis et al., 2016). It is the responsibility of the public health officials to ensure the population is protected from occurring or reoccurring health challenges and to mitigate the effects if and when these health issues occur (IOM, 1988).

Figure 6.1 shows, in a nutshell, the typical components of public health systems, from hospitals to laboratories, to faith-based institutions to drug treatment and rehabilitation outfits to work places, to homes etc. involving partnerships and communication that are all intertwined in the system.

6.5 CLASSIFICATION OF PUBLIC HEALTHCARE SYSTEMS

Jarvis et al. (2016) have classified public healthcare systems into financial arrangements, delivery arrangements, and governance arrangements.

6.5.1 Financial Arrangements

According to Wiysonge et al. (2014), this has to do with finding new ways to collect and allocate funds for healthcare delivery, how services are paid for, targeted incentives and disincentives and insurance schemes. Health systems has as a core function health financing. This has an overbearing effect on the progress towards universal health coverage which will in turn improve effective service delivery and financial coverage. The cost of public healthcare today makes it difficult for millions of people to access public healthcare. It's normal to expect people to pay for medical care out-of-pocket yet with unsatisfactory services most times. Robust carefully designed and executed financial arrangements can help fix these issues for instance, the involvement of contracting agents to handle payments. In which case, disbursement of funds to providers will receive the urgency that it should be treated with. This will also create space for adequate staffing and consumables for treatment (WHO, 2022).

6.5.2 Delivery Arrangements

Delivery arrangements is another class of healthcare system. This has to do with changes in who will be receiving the care and when, the working conditions of those providing the care, the person to administer care to patients, use of information technology for health information dissemination and coordination of care among different providers. Quality and safety systems too are part of this (Ciapponi et al., 2017). Mode of service delivery impacts greatly on effectiveness, efficiency, and equity of health systems. HMOs were introduced into the PHC system to help ameliorate healthcare delivery. An HMO is a NHIS, National Health Insurance Scheme, registered private or public incorporated company that provides and manages PHC services through the accredited healthcare facilities as appointed by the Health Insurance Scheme. However, as at 2021, there were only about 60 HMOs in Nigeria. This number is grossly inadequate to serve the citizenry (Kuo, 2020). Some accredited HMOs in Nigeria are Total Health Trust, which offers quality health insurance services to individuals, small businesses and large corporate organizations, United Healthcare International, Prepaid Medical Services, and a host of others.

6.5.3 Governance Arrangements

Any changes that will bring about accountability and authority on health policies, commercial products, health organizations, and health professionals by the government but also involving stake holders in decision-making is termed Governance Arrangement. The government funds and maintains public healthcare structure of their citizens. Sadly, this is not the case in most African countries especially Nigeria. Despite the 2001 Abuja Declaration by African heads of states where it was agreed that at least 15% of the countries' national budget will be allocated to healthcare, Nigerian governments have continued to allocate less than 10% annually. Maternal mortality in Nigeria ranked 20% globally in 2019. From research findings, there were just about 24,000 hospitals across the whole country.

6.6 ORGANIZATIONS IN PHS

Some of the sectors and organizations that are involved in the PHS include:

- Public health agencies. These are agencies set up by the government at local or state levels to serve as governments' representatives to create and ensure the existence of a functional public health system.

- Private entities like hospitals, clinics, community health centers, mental health rehabilitation centers, medical laboratories, and nursing homes for the aged, which provide care that prevents, cures, and rehabilitates.
- Organizations concerned with public safety like the police departments, fire service, and emergency medical services, whose responsibility it is to prevent and cope with injuries and the injured.
- Human charity and human service organizations. These are concerned with assisting people to receive healthcare and other health enhancing services. They include transportation service providers, food banks, and public assistance agencies.
- Educational and youth development organizations. These are schools, faith-based organizations, youth centres, vocational centres whose concern is to educate, inform and prepare children and young people to make informed decisions concerning their health, life choices and to make conscious efforts to be useful to their families and society
- Arts-related and recreational organizations. These are places open to the community where the mental and physical health of members of the community can be taken care of. Economic organizations, and philanthropic organizations all provide the right structures and resources for individuals, families and organizations to survive and thrive in the community. These are zoning boards, community development organizations, and employers of labor and business foundations.
- Government or private environmental agencies also contribute to make policies and enforce laws that relate to or advocate also for a healthy environment.

6.7 MODELS FOR PUBLIC HEALTHCARE SYSTEMS

Four major models have been proposed for PHS: the Beveridge Model, the Bismarck Model, the National Health Insurance model, and the out-of-pocket models. In theory, these four categories are distinct but usually there is a blend of all four in most countries.

6.7.1 The Beveridge Model

Sir William Beveridge developed this model in 1948. Often, a National Health Service is established for centralizing the system. The United Kingdom government as the single player in the global market, eliminates competition and keeps all prices low. This model is also seen to be adopted in many areas of Northern Europe and in several other parts of the world. In this model, income taxes are used to fund healthcare. Hence, there are free services for all medical emergencies and interventions. As long as a patient is a tax payer, then they don't have to pay for their medical services. The majority of health staff in this model are government employees and health is seen as a human right. Hence all citizens are entitled to the free healthcare services. The only downside with this model is the long waiting lists. Also, there's the tendency for the system to be over utilized since it is free, causing a gradual wear and tear that will impact negatively on the cost. This fear was rife in America where it was believed that adopting this model may lead to unnecessary increase in demand for healthcare services, even procedures that are not necessary in any way just because it is free.

Also in the face of a national crisis where government resources need to be channeled to solve other problems, public health services funding may drop drastically as public revenues also reduce.

In the 19th century, however, a more decentralized healthcare model was introduced by Otto von Bismarck. He called it the Bismarck Model. In this model, both employees and employers fund health insurance and deductions are made from source.

It is a model that focuses resources on those who can contribute financially because it was not established to provide universal coverage. The funding in this model is seen as public but the health providers are usually private entities. There could be single insurers, multiple/competing insurers or even multiple/non-competing insurers. Here the government tightly controls prices, irrespective of the number of insurers and the insurers are not out for profit-making.

This model ensures that the workers will have adequate healthcare that will make them continue working to ensure a productive workforce.

This model, however, has some concerns such as how to care for those unable to work or those who may not be able to afford contributions. Other concerns are how to contend with aging populations, with an uneven number of retired citizens compared to employed citizens, and how to stay competitive in attracting international companies that may prefer locations without these required payroll dedications. The shift in mindset from seeing health as a privilege for only the privileged to seeing health as a right for all citizens, highlights these concerns.

6.7.2 The National Health Insurance Model

This model incorporates aspects of both the Bismarck and Beveridge models. There's still a single payer for medical procedures, which is the government in partnership with private providers. This model does not deny claims and also does not make profits off of citizens. Many countries seem to adopt a hybrid model that incorporates both models, like Hungary and Germany. In Canada, insurance is still private but optional. In Nigeria and other African countries, the federal government also offers this universal model to government parastatals that will sign up for it. It is important to note that hospitals still maintain their independence. They also use insurance policies to reduce internal complications since there's a balance between private practice and public insurance. Patients choose their private providers because barriers to treatment are usually low. Regardless of income level, just like the Beveridge Model, most procedures are included in this system. Since the government has to process all claims, it reduces the amount of duplication of services, hence, the cost of administration of health insurance is drastically reduced. One disadvantage worth noting is the wait time for implementation. It is considered a serious health policy issue.

6.7.3 The Out-of-Pocket Model

This model, also referred to as a 'market-driven healthcare' is popular in less developed countries as there are too few resources to cater for the medical needs of the citizenry. Hence, citizens have to pay for medical services from out-of-pocket. Cash, in this case, determines who can receive the best care. This is the situation in most African countries including Nigeria. No money, no proper healthcare services.

In 2017 the World Health Organization (WHO) reported that 77% of Nigerians pay for their healthcare services from their own pocket, meaning that most Nigerians may never have heard of neither do they have any form of health insurance. Apparently, the poorer you are, the less access to quality healthcare to which you are entitled. Okunola (2020) stated that "Policy makers and political actors need to stop the high reliance on out-of-pocket health care spending as a means of financing the Nigerian health system by increasing public health expenditure."

However it's not only in African countries that health insurance is a problem. It is on record that socioeconomic status and ethnicity affects healthcare in a country like America more than in Canada. The percentage of persons without health insurance in 2015 was 13%. There's still an ongoing debate to increase coverage and minimize the cost of healthcare in the American congress (Princeton Public Health Review, 2017).

6.8 FUNCTIONS OF PUBLIC HEALTH SYSTEMS

- It is the duty of the public health system to solve community health issues by monitoring citizens health status.
- If there are any community health hazards issues, the public health system should investigate and diagnose these health problems.
- The public health system also gives information about health issues, and educates and empowers the people about their health.
- If there's a need to mobilize a community-based partnership and action, either to identify or solve any health problem, the PHS handles this.
- Policies, plans, and programs that support health efforts by individuals and communities are also developed by PHS.
- Regulations and laws that protect health and safety are also enforced by the PHS.
- PHS also gets people to know where there is needed health services and provide assurances of healthcare if unavailable immediately.
- This system ensures there is competent and effective healthcare workforce both at the public and private levels.
- The accessibility, effectiveness, and quality of personal and community-based health services, are evaluated by PHS.
- Very interestingly, they also carry out research on new health challenges and innovative solutions to these health solutions.

There's so much happening in the healthcare system in recent times that it is changing the way PHS work and prompting novel roles for PH agencies. Some of these are a growing focus on integrating public health and primary healthcare, an examination of financing options for core foundational services and new ways of delivering services more efficiently and effectively, for instance, with cross-jurisdictional sharing, which can involve sharing defined services like laboratory testing, and informal collaborations that may need support to formal structural changes like consolidating public health agencies.

6.9 CONCLUSION

Nigeria as a country with insecurity, political and economic issues, may not be ready to adapt personalized healthcare into the national healthcare scheme, even though there are plans to do so by 2030. Long after independence, the country adopted a welfare-based system for its public healthcare delivery where the government offers free or heavily subsidized treatment and medical services. In recent times, however, it has become difficult for the government to continue.

WHO's health system evaluation and ranking of Nigeria may have improved from 187 out of 191 to 163, over two decades, but it's still one of the lowest in the world.

Lack of qualified workers in the health sector seems to be one of our greatest challenges as most of Nigeria's best healthcare service experts have migrated out of the country to greener pastures in recent times. On April 24, 2020, Chris Ngige, Nigeria's minister of labor, who happens to be a medical doctor himself, made a sweeping statement that Nigeria has "more than enough" doctors. But this is erroneous because, in reality, there are only about 0.38 doctors per 1,000 people in the country (Okunola, 2020).

Poor remuneration and welfare packages are rife in the health sector as in several other sectors and this also affects healthcare delivery in Nigeria. Records show that Nigerian healthcare statistics are abysmal. Life expectancy is recorded to be as low as 54.4%. With a recorded global maternal mortality rate at 19%, Nigeria has one of the worst recorded global maternal mortality rates in the world. On the other hand, an infant mortality rate of 19 deaths for every 1,000 births is recorded and children under 5 record a mortality rate of 128 per 1,000.

As earlier mentioned, Nigerian out-of-pocket healthcare expenditure is a whopping 71.7% – expenses that are not paid for government subsidies or insurance. As at 2016, only 3.7% of Nigeria's GDP could be attributed to healthcare services. Inequality has also been identified as a major role player in plaguing the healthcare services delivery in Nigeria. Healthcare services are more easily accessible in urban areas than in the rural villages. There have been reports too that in some regions of the country, health workers out on campaigns are antagonized especially when there are unfounded reports of deadly vaccine propaganda flying around in social media. However, some of these setbacks have been recognized and the country is trying out different ways to mitigate them. One such is to pay health workers an extra allowance to relocate to the rural areas. Another is the midwifery service scheme where recently graduated, unemployed or retired midwives are employed and placed in rural areas for one year's community service. This intervention seems to be very successful so far (Kuo, 2020).

In recent times, Nigeria has stood out among nations in tackling disease outbreaks successfully, successes that can be attributed more to the healthcare givers than to the government. A drive to end polio saw a nationwide campaign being launched and executed judiciously in different states by dedicated healthcare workers moving literally from house to house and door to door to ensure infants and children are vaccinated. During the break out of Ebola, Lassa fever, and recently COVID-19, healthcare services were so organized to successfully control and curb these outbreaks. Some of these wins can be attributed to strategies like better vaccine storage methods and decentralized disease control network.

If the low points like healthcare inequality and inadequate qualified healthcare workforce (consider the recent emigration of health workers to Saudi Arabia) can be arrested, then the country will be on the way out of the healthcare crisis. Modernizing the healthcare infrastructure will also go a long way to place the country at par with other African countries.

With new insights and breakthroughs in the world of medicine coming through from across the world, the health status of billions of people is radically improving. Universities, civil societies, policymakers and enterprises/pharmaceuticals are ensuring these changes in medicine and public healthcare. Disease outbreak, bioattack, and other life-threatening incidences have been controlled by automated medical intelligence and surveillance. Access to GPS systems also makes it easy to monitor areas that would have otherwise been difficult to access.

MIS systems need to be made a part of the Nigerian healthcare system for record keeping from hospital visits to drug delivery and even clean bills of health, to serve the needs of the healthcare systems of this modern era.

REFERENCES

Bresnick, J. (2018). Top 12 ways artificial intelligence will impact healthcare. *Health IT Analytics*, April 30, 2018. Retrieved at https://healthitanalytics.com/news/top-12-ways-artificial-intelligence-will-impact-healthcare

Ciapponi A, Lewin S, Herrera CA, Opiyo N, Pantoja T, Paulsen E, Rada G, Wiysonge CS, Bastías G, Dudley L, Flottorp S, Gagnon M-P, Marti SG, Glenton C, Okwundu CI, Peñaloza B, Suleman F, Oxman AD, Cochrane Effective Practice and Organisation of Care Group (2017). Delivery arrangements for health systems in low-income countries: An overview of systematic reviews. *Cochrane Database of Systematic Reviews*. doi: 10.1002/14651858.CD011083.pub2

Edward-Ekpu, U. (2014) Two Nigerian laboratories have taken big steps to boost genetics medicine in Africa. https://qz.com/africa/1945960/nigerian-labs-acegid-54gene-boost-genetics-medicine-in-africa/. Retrieved on 2022, March 29.

Friedman N, Shemer J, Kokia E (2003) Demonstrating value-added utilization of existing databases for organizational decision–support. *The Israel Medical Association Journal* 5: 3–8. doi: 10.4018/irmj.2002100101

Gatera M, Uwimana J, Manzi E, Ngabo F Nwaigwe F Gessner BD, Moïsi JC (2016) Use of administrative records to assess pneumococcal conjugate vaccine impact on pediatric meningitis and pneumonia hospitalizations in Rwanda. *J. Vaccine*. 34(44): 5321–5328. doi: 10.1016/j.vaccine.2016.08.084

Goshen R, Mizrahi B, Akiva P, Kinar Y, Choman E, Shalev V, et al.(2017) Predicting the presence of colon cancer in members of a health maintenance organisation by evaluating analytes from standard laboratory records. *British Journal of Cancer* 116: s944–s950. doi: 10.1038/bjc.2017.53

IOM (Institute of Medicine). (1988) *The Future of Public Health*. Washington, DC: National Academy Press. [PubMed]

Jarvis T, Scott F, El-Jardali F, Alvarez E (2016) Defining and classifying public health systems: A critical interpretive synthesis. *Health Research Policy and Systems* 18: 68.

Kuo E (2020) https://borgenproject.org/healthcare-in-nigeria/

Lewy H, Barkan R, Sela T (2019) Personalized health systems—Past, present, and future of research development and implementation in real-life. *Frontiers in Medicine*. doi: 10.3389/fmed.2019.00149

Medical Tourism Magazine (2022) www.magazine.medicaltourism.com/article/precision-medicine-reshaping-healthcare-europe

Okunola, A. (2020) www.globalcitizen.org/en/content/health-care-facts-nigeria-covid-19/. Retrieved on 2022, March 23.

Princeton Public Health Review (2017) https://pphr.princeton.edu/2017/12/02/unhealthy-health-care-a-cursory-overview-of-major-health-care-systems/. Retrieved on 2022, March 30.

World Economic Forum (WEF) (2022) 3 ways China is leading the way in precision medicine. www.weforum.org/agenda/2017/11/3-ways-china-is-leading-the-way-in-precision-medicine/

WHO (2022) Health financing. www.who.int/health-topics/health-financing#tab=tab_1

World Health Organization. MHealth (2016) Use of mobile wireless technologies for public health (Report EB139/8). https://apps.who.int/gb/ebwha/pdf_files/EB139/B139_8-en.pdf. Retrieved on 2022, February 15.

Wiysonge CS, Herrera CA, Ciapponi A, Lewin S, Garcia Marti S, Opiyo N, et al. (2014) Financial arrangements for health systems in low-income countries: An overview of systematic reviews. *Cochrane Database of Systematic Reviews*. doi: 10.1002/14651858.CD011084

7 Application of Artificial Intelligence in Dentistry and Orthodontics

Igiku Victory, Charles Oluwaseun Adetunji, and Akinola Samson Olayinka
Edo State University, Uzairue, Iyamho, Nigeria

Olugbemi T. Olaniyan
Rhema University, Aba, Nigeria

Juliana Bunmi Adetunji
Osun State University, Osogbo, Nigeria

Tosin Comfort Olayinka
Wellspring University, Benin City, Nigeria

Aishatu Idris Habib
Edo State University, Uzairue, Iyamho, Nigeria

Olorunsola Adeyomoye
University of Medical Sciences, Ondo City, Nigeria

Oluwafemi Adebayo Oyewole
Federal University of Technology, Minna, Nigeria

K. I. T. Eniola
Joseph Ayo Babalola University, Ikeji Arakeji, Nigeria

 DOI: 10.1201/9781003309468-7

7.1 INTRODUCTION

The main scope of artificial intelligence appliances in medicine includes diagnostics and prognostics, decision-making, and robotics all supported by artificial intelligence algorithms such as artificial neural networks (ANN) (Ramesh et al. 2004), expert systems (Seto et al. 2012), precision control, and automation (Ashrafian et al. 2014).

The introduction of artificial intelligence in dentistry and orthodontics has revolutionized the field, offering unprecedented possibilities for enhancing patient care and streamlining clinical workflows. Its applications range from improving the accuracy of diagnoses to optimizing treatment plans and coordinating appointments, making the practice more efficient and effective. Artificial intelligence can also notify the dentist about the medical history of patients and also patient habits such as smoking, alcoholism, and diet.

Recent advances in ANNs can detect and diagnose lesions which are unseen by the human eye. With artificial intelligence technology, patients can be notified of the most probable diagnosis of an illness, by entering their symptoms on to their smart devices (Bhatia et al., 2019). Through the help of artificial intelligence technologies, dental students can undergo their preclinical work on virtual patients instead of simulations or phantom hands (Yau et al., 2006). This reduces the risk associated with working on a live patient (Murray, 2020).

Artificial intelligence is very important in every aspect of medical science (Adetunji et al., 2020; Adetunji et al., 2021; Oyedara et al., 2022; Adetunji et al., 2022a,b,c,d,e,f,g,h,Adetunji and Oyeyemi, 2022; Olaniyan et al., 2022a, b). Hamet and Tremblay (2017), divided the application of artificial intelligence in medicine into two categories: (1) the virtual application, which ranges from the use of artificial intelligence in electronic health records systems, to neural network-based guidance in treatment decision and (2) the physical application such as intelligent prosthesis for handicapped people and elderly care and robotics.

The evolution of robotic surgery, in which artificial intelligence imitates the simulation of human body motion and intellect, is the principal application of artificial intelligence in dental surgery (Humairo et al., 2021). Oral implant surgery, tumor and foreign body removal, biopsy, and temporomandibular joint surgery are all successful clinical applications of artificial intelligence in image-guided surgery in the cranial areas (Bindushree et al. 2022).

7.2 ROLE OF ARTIFICIAL INTELLIGENCE IN DENTISTRY

Artificial intelligence can be applied in the diagnosis and treatment of patients by dentists through the evaluation of large sets of clinical data. Images of lesions can be uploaded to an artificial intelligence system and diagnosis can be made by running the image against a large set of data (Bhatia et al., 2019). Neural networks are significant in the diagnosis and treatment of dental diseases especially in the diagnosis of diseases without a specific or no etiological cause (Jain and Wynne, 2021).

In a study by Dar-Odeh et al. (2010), data from 86 participants were used to build and train an ANN model to predict the factors responsible for recurring aphthous ulcers. The factors identified to be responsible for recurring aphthous ulcers were gender, hemoglobin, serum vitamin B12, serum ferritin, red cell folate, salivary candida colony count, frequency of tooth brushing and number of fruits or vegetables consumed per day. Artificial intelligence has also been used in the detection of dental caries which is one of the common dental diseases. Geetha et al. (2020), proposed an ANN algorithm for the diagnosis of dental caries using 105 radiograph images. The 16-feature vector extracted from the segmented images was used by the authors as input nodes, the output nodes consisted of caries or sound tooth. The model achieved an accuracy of 97.1% on the detection of caries and had a false negative of 2.8%, indicating that the neural network could predict the caries more accurately than conventional dental examinations.

In endodontics, artificial intelligence can be used to detect vertical lesions and root fractures, on cone beam computed tomography (CBCT) images and radiographs, anatomy of the root canal evaluation, predicting the viability of the dental pulp stem cells, predict the success of retreatment procedures (Aminoshariae et al. 2020). Johari et al. (2017), developed a probabilistic neural network (PNN) model for the diagnosis of vertical root fractures on cone beam computed tomography images of endodontically treated and intact teeth. The model displayed an excellent performance with an accuracy of 96.6% in the diagnosis of vertical root fractures.

In a study by Setzer et al. (2020), the authors used a deep learning algorithm for the detection of periapical lesion on CBCT images. The algorithm achieved an accuracy of 93% in the detection of the lesions. Similarly, Orhan et al. (2020), detected periapical lesions on CBCT images using a convolutional neural network (CNN). The results obtained in this study were similar to those obtained by a dental practitioner.

Several researchers have also proposed the use of artificial intelligence technology in periodontics. Krois et al. (2019), used a deep CNN to detect periodontal bone loss on panoramic dental radiographs. The CNN achieved an accuracy of 83%. Dental periapical radiographs can be used to detect peri-implant bone loss; the downside of this is that the boundaries of bone around the implants are often unclear or may overlap (Ossowska et al. 2022). The advantage of CNNs in this context is that they can assess the marginal bone level, top and apex of the implant on dental periapical radiographs (Ossowska et al. 2022).

Jun-Young Cha et al. (2021), proposed an automated assistant system to calculate the bone loss percentage of peri-implantitis and classify the resorption severity of peri-implantitis. Lee et al. (2018), developed a deep CNN model for the diagnosis and prediction of periodontally compromised teeth using 64 premolars and 64 molars clinically diagnosed of severe periodontitis. The model achieved a diagnostic accuracy of 81% for premolar and 76.7% for molars.

Yauney et al. (2019), reported an automated process based on a CNN that correlated poor periodontal health to systemic health symptoms. The authors concluded that the model could be used for the diagnosis and systemic health screening of other diseases. Rana et al. (2021), proposed a framework that successfully distinguishes between inflamed and healthy gingiva.

Artificial intelligence can also be used in dental surgeries. One important application of artificial intelligence in this aspect is in image-guided surgery, which provides a more accurate surgical resection, decreasing the need for reversion procedures. Patcas et al. (2019), reported the use of artificial intelligence to describe the impact of orthognatic treatment on facial attractiveness and age appearance. In total 146 pre- and post-treatment photographs of orthognatic patients were collected. The algorithm showed that the appearance of most patients improved with treatment (66.4%).

CNNs may be used to identify dental implant brands on panoramic radiographs and to identify the stage of treatment (Sukegawa et al. 2021). The application of artificial intelligence in dental surgeries includes areas such as orthognatic surgeries, changes in bone loss or post extraction complications, and implantology treatments (Ossowska et al. 2022).

Artificial intelligence virtual dental assistants can perform different tasks in dental clinics with less error and more precision. Artificial intelligence can be applied in booking dental appointments, taking medical and dental history of the patient, managing insurance as well as assisting the dental surgeon in adequate diagnosis. Artificial intelligence may also be used to predict the genetic predisposition of a large population to oral cancer (Majumdar et al. 2018). It can also be used to assess the quality of osteo-integration (Ossowska et al. 2022).

7.3 APPLICATION OF ARTIFICIAL INTELLIGENCE IN ORTHODONTICS

The application of artificial intelligence is also growing in the unique field of orthodontics. Accurate diagnosis, treatment planning, and prognosis prediction are all essential components of successful orthodontic therapy (Khanagar et al. 2021). The commonly used artificial intelligence algorithm in this area is the CNN, ANN, support vector machine, and regression algorithms (Bichu et al. 2021).

Artificial intelligence can be used in orthodontics for the diagnosis and treatment planning through the analysis of radiographs and photographs by intra-oral scanners and cameras which removes the need for making patients impressions and several laboratory steps (Sharma, 2019). Xie et al. (2010), developed an expert decision-making model based on an ANN, for the orthodontic treatment of patients aged between 11 to 15 years old, to determine whether tooth extraction is necessary. The model achieved an accuracy of 80% in testing the set. Artificial intelligence can also be used to provide orthodontic consultations to general practitioners for the alignment of crowded lower teeth (Kalappana et al., 2018). Similarly, Pellini et al. (2020), predicted whether a patient needed extractions or not in their treatment plan using an ANN. The model showed an excellent result in both the accuracy of extraction and prediction of the use of maximum anchorage.

Artificial intelligence technologies can be used in cephalometric tracing to save time by reducing the tracking errors and increasing the diagnostic value of cephalometric analysis (Yu et al., 2020). Lee et al. (2018), reported the use of deep convolutional neural network (DCNN) based analysis for cephalometric tracing. The study showed that the model had an accuracy of 90% in the identification of cephalometric landmarks. Muraev et al. (2020) reported the use of an ANN for cephalometric tracing on cephalometric radiographs. The authors compared the accuracy of cephalometric tracing between the ANN model and three groups of doctors: expert, regular, and inexperienced. The study showed that the ANN model had a similar accuracy in cephalometric tracing to that of the inexperienced dentist.

7.4 CONCLUSION

The chapter shows that artificial intelligence could be a major part of modern medicine. The application of artificial intelligence in dentistry includes patient diagnosis and treatment, artificial intelligence-assisted decision-making, and artificial intelligence-assisted dental surgeries. Advances in artificial intelligence especially in its application in dentistry could be of tremendous importance to dentists and researchers in the improvement of patient healthcare.

REFERENCES

Adetunji, Charles Oluwaseun, Egbuna, Chukwuebuka, Oladosun, Tolulope Olawumi, Akram, Muhammad, Michael, Olugbenga, Olisaka, Frances Ngozi, Ozolua, Phebean, Adetunji, Juliana Bunmi, Enoyoze, Goddidit Esiro, & Olaniyan, Olugbemi (2021). Efficacy of phytochemicals of medicinal plants for the treatment of human echinococcosis. In *Neglected tropical diseases and phytochemicals in drug discovery*. DOI: 10.1002/9781119617143.ch8

Adetunji, Charles Oluwaseun, Mitembo, William Peter, Egbuna, Chukwuebuka, Narasimha Rao, G.M. (2020). In silico modeling as a tool to predict and characterize plant toxicity. In Andrew G. Mtewa, Chukwuebuka Egbuna, & G.M. Narasimha Rao (Eds.), *Poisonous plants and phytochemicals in drug discovery*. Wiley Online Library DOI: 10.1002/9781119650034.ch14

Adetunji, Charles Oluwaseun, Nwankwo, Wilson, Olayinka, Akinola Samson, Olugbemi, Olaniyan Tope, Akram, Muhammad, Laila, Umme, Olugbenga, Michael Samuel, Oshinjo, Ayomide Michael, Adetunji, Juliana Bunmi, Okotie, Gloria E., & Esiobu, Nwadiuto (Diuto) (2022h). Machine learning and behaviour modification for COVID-19. In *Medical biotechnology, biopharmaceutics, forensic science and bioinformatics* (1st ed.). First Published 2022. Imprint CRC Press. Pages 17. eBook ISBN 9781003178903. DOI: 10.1201/9781003178903-17

Adetunji, Charles Oluwaseun, Nwankwo, Wilson, Olayinka, Akinola Samson, Olugbemi, Olaniyan Tope, Akram, Muhammad, Laila, Umme, Samuel, Michael Olugbenga, Oshinjo, Ayomide Michael, Adetunji, Juliana Bunmi, Okotie, Gloria E., & Esiobu, Nwadiuto (Diuto) (2022f). Computational intelligence techniques for combating COVID-19. In *Medical biotechnology, biopharmaceutics, forensic science and bioinformatics* (1st ed.). First Published 2022. Imprint CRC Press. Pages 12. eBook ISBN 9781003178903. DOI: 10.1201/9781003178903-16

Adetunji, Charles Oluwaseun, Olaniyan, Olugbemi Tope, Adeyomoye, Olorunsola, Dare, Ayobami, Adeniyi, Mayowa J, Alex, Enoch, Rebezov, Maksim, Garipova, Larisa, & Shariati, Mohammad Ali. (2022a). eHealth, mHealth, and telemedicine for COVID-19 pandemic. In S.K. Pani, S. Dash, W.P. dos Santos, S.A. Chan Bukhari, & F. Flammini (Eds.), *Assessing COVID-19 and other pandemics and epidemics using computational modelling and data analysis*. Springer. DOI: 10.1007/978-3-030-79753-9_10

Adetunji, Charles Oluwaseun, Olaniyan, Olugbemi Tope, Adeyomoye, Olorunsola, Dare, Ayobami, Adeniyi, Mayowa J., Alex, Enoch, Rebezov, Maksim, Isabekova, Olga, & Shariati, Mohammad Ali (2022c). Smart sensing for COVID-19 pandemic. In S.K. Pani, S. Dash, W.P. dos Santos, S.A. Chan Bukhari, & F. Flammini (Eds.), *Assessing COVID-19 and other pandemics and epidemics using computational modelling and data analysis*. Springer. DOI: 10.1007/978-3-030-79753-9_9

Adetunji, Charles Oluwaseun, Olaniyan, Olugbemi Tope, Adeyomoye, Olorunsola, Dare, Ayobami, Adeniyi, Mayowa J., Alex, Enoch, Rebezov, Maksim, Koriagina, Natalia, & Shariati, Mohammad Ali (2022e). Diverse techniques applied for effective diagnosis of

COVID-19. In S.K. Pani, S. Dash, W.P. dos Santos, S.A. Chan Bukhari, & F. Flammini (Eds.), *Assessing COVID-19 and other pandemics and epidemics using computational modelling and data analysis*. Springer. DOI: 10.1007/978-3-030-79753-9_3

Adetunji, Charles Oluwaseun, Olaniyan, Olugbemi Tope, Olorunsola Adeyomoye, Ayobami Dare, Mayowa J. Adeniyi, Enoch Alex, Maksim Rebezov, Ekaterina Petukhova, & Shariati, Mohammad Ali (2022b). Machine learning approaches for COVID-19 pandemic. In S.K. Pani, S. Dash, W.P. dos Santos, S.A. Chan Bukhari, & F. Flammini (Eds.), *Assessing COVID-19 and other pandemics and epidemics using computational modelling and data analysis*. Springer. DOI: 10.1007/978-3-030-79753-9_8

Adetunji, Charles Oluwaseun, Olaniyan, Olugbemi Tope, Adeyomoye, Olorunsola, Dare, Ayobami, Adeniyi, Mayowa J., Alex, Enoch, Rebezov, Maksim, Petukhova, Ekaterina, & Shariati, Mohammad Ali (2022d). Internet of Health Things (IoHT) for COVID-19. In S.K. Pani, S. Dash, W.P. dos Santos, S.A. Chan Bukhari, & F. Flammini (Eds.), *Assessing COVID-19 and other pandemics and epidemics using computational modelling and data analysis*. Springer. DOI: 10.1007/978-3-030-79753-9_5

Adetunji, Charles Oluwaseun, Olugbemi, Olaniyan Tope, Akram, Muhammad, Laila, Umme, Samuel, Michael Olugbenga, Oshinjo, Ayomide Michael, Adetunji, Juliana Bunmi, Okotie, Gloria E., Esiobu, Nwadiuto (Diuto), Oyedara, Omotayo Opemipo, & Adeyemi, Folasade Muibat (2022g). Application of computational and bioinformatics techniques in drug repurposing for effective development of potential drug candidate for the management of COVID-19. In *Medical biotechnology, biopharmaceutics, forensic science and bioinformatics* (1st ed.). First Published 2022, Imprint CRC Press. Pages 14. eBook ISBN 9781003178903. DOI: 10.1201/9781003178903-15

Adetunji, Charles Oluwaseun, & Oyeyemi, Oyetunde T. (2022). Antiprotozoal activity of some medicinal plants against entamoeba histolytica, the causative agent of amoebiasis. In *Medical biotechnology, biopharmaceutics, forensic science and bioinformatics* (1st ed.). First Published 2022. Imprint CRC Press. Pages 12. eBook ISBN 9781003178903. https://www.taylorfrancis.com/chapters/edit/10.1201/9781003178903-20/antiprotozoal-activity-medicinal-plants-entamoeba-histolytica-causative-agent-amoebiasis-charles-oluwaseun-adetunji-oyetunde-oyeyemi

Aminoshariae, A., Kulild, J., & Gutmann, J. (2020). The association between smoking and periapical periodontitis: A systematic review. *Clinical Oral Investigations*, *24*(2), 533–545. https://doi.org/10.1007/s00784-019-03094-6

Ashrafian, H., Toma, T., Harling, L., Kerr, K., Athanasiou, T., & Darzi, A. (2014). Social networking strategies that aim to reduce obesity have achieved significant although modest results. *Health Affairs*, *33*(9), 1641–1647. https://doi.org/10.1377/hlthaff.2014.0370

Bhatia, S. K., Tiwari, S., Mishra, K. K., & Trivedi, M. C. (Eds.). (2019). *Advances in Computer Communication and Computational Sciences: Proceedings of IC4S 2018* (Vol. 924). Springer Singapore. https://doi.org/10.1007/978-981-13-6861-5

Bichu, Y. M., Hansa, I., Bichu, A. Y., Premjani, P., Flores-Mir, C., & Vaid, N. R. (2021). Applications of artificial intelligence and machine learning in orthodontics: A scoping review.*Progress in Orthodontics*, *22*(1), 18. https://doi.org/10.1186/s40510-021-00361-9

Bindushree, et al. (2022). *Artificial intelligence: In modern dentistry*. Retrieved May 28, 2022, from https://www.jdrr.org/article.asp?issn=23482915;year=2020;volume=7;issue=1;spage=27;epage=31;aulast=Bindushree

Cha, D., Pae, C., Lee, S. A., Na, G., Hur, Y. K., Lee, H. Y., Cho, A. R., Cho, Y. J., Han, S. G., & Kim, S. H. (2021). Differential biases and variabilities of deep learning–based artificial intelligence and human experts in clinical diagnosis: Retrospective cohort and survey study. *JMIR Medical Informatics*, *9*(12), e33049.

Dar-Odeh, N. S., Alsmadi, O. M., Bakri, F., Abu-Hammour, Z., Shehabi, A. A., Al-Omiri, M. K., AbuHammad, S. M. K., Al-Mashni, H., Saeed, M. B., Muqbil, W., & Abu-Hammad, O. A. (2010). Predicting recurrent aphthous ulceration using genetic algorithms-optimized neural networks. *Advances and Applications in Bioinformatics and Chemistry: AABC*, *3*, 7–13.

Geetha, V., Aprameya, K. S., & Hinduja, D. M. (2020). Dental caries diagnosis in digital radiographs using back-propagation neural network. *Health Information Science and Systems*, *8*(1), 8. https://doi.org/10.1007/s13755-019-0096-y

Hamet, P., & Tremblay, J. (2017). Artificial intelligence in medicine. *Metabolism*, *69*, S36–S40. https://doi.org/10.1016/j.metabol.2017.01.011

Humairo, C. N., Hapsari, A., & Bramanti, I. (2021). The role of artificial intelligence in many dental specialties. *Bio Web of Conferences*, *41*, 03005. https://doi.org/10.1051/bioconf/20214103005

Jain, P., & Wynne, C. (2021). Artificial intelligence and big data in dentistry. In P. Jain & M. Gupta (Eds.), *Digitization in dentistry: Clinical applications* (pp. 1–28). Springer International Publishing. https://doi.org/10.1007/978-3-030-65169-5_1

Johari, M., Esmaeili, F., Andalib, A., Garjani, S., & Saberkari, H. (2017). Detection of vertical root fractures in intact and endodontically treated premolar teeth by designing a probabilistic neural network: An ex vivo study. *Dentomaxillofacial Radiology*, *46*(2), 20160107. https://doi.org/10.1259/dmfr.20160107

Kalappanar, A., Sneha, S., & Rajeshwari, G. A. (2018). *Artificial intelligence: A dentist's perspective - ProQuest*. Retrieved May 28, 2022, from https://www.proquest.com/openview/29cd302ceda870913572433757aef4c9/pq-origsite=gscholar&cbl=2068935

Khanagar, S. B., Al-Ehaideb, A., Maganur, P. C., Vishwanathaiah, S., Patil, S., Baeshen, H. A., Sarode, S. C., & Bhandi, S. (2021). Developments, application, and performance of artificial intelligence in dentistry – A systematic review. *Journal of Dental Sciences*, *16*(1), 508–522. https://doi.org/10.1016/j.jds.2020.06.019

Krois, J., Graetz, C., Holtfreter, B., Brinkmann, P., Kocher, T., & Schwendicke, F. (2019). Evaluating Modeling and Validation Strategies for Tooth Loss. *Journal of Dental Research*, *98*(10), 1088–1095. https://doi.org/10.1177/0022034519864889

Lee, J.-H., Kim, D.-H., Jeong, S.-N., & Choi, S.-H. (2018). Detection and diagnosis of dental caries using a deep learning-based convolutional neural network algorithm. *Journal of Dentistry*, *77*, 106–111. https://doi.org/10.1016/j.jdent.2018.07.015

Monill-González, A., Rovira-Calatayud, L., d'Oliveira, N. G., & Ustrell-Torrent, J. M. (2021). Artificial intelligence in orthodontics: Where are we now? A scoping review. *Orthodontics & Craniofacial Research*, *24*(S2), 6–15. https://doi.org/10.1111/ocr.12517

Muraev, A. A., Tsai, P., Kibardin, I., Oborotistov, N., Shirayeva, T., Ivanov, S., Ivanov, S., Guseynov, N., Aleshina, O., Bosykh, Y., Safyanova, E., Andreischev, A., Rudoman, S., Dolgalev, A., Matyuta, M., Karagodsky, V., & Tuturov, N. (2020). Frontal cephalometric landmarking: Humans vs artificial neural networks. *International Journal of Computerized Dentistry*, *23*(2), 139–148.

Murray, D. (2020). *Using Human Rights Law to Inform States' Decisions to Deploy AI*. https://www.cambridge.org/core/journals/american-journal-of-international-law/article/using-human-rights-law-to-inform-states-decisions-to-deploy-ai/44C4808E7E42F172E0125497CF1096E2

Olaniyan, Olugbemi T., Adetunji, Charles O., Adeniyi, Mayowa J., & Hefft, Daniel Ingo (2022a). Machine learning techniques for high-performance Computing for IoT applications in healthcare. In *Deep learning, machine learning and iot in biomedical and health informatics* (1st ed.). First Published 2022. Imprint CRC Press. Pages 13. eBook ISBN 9780367548445. DOI: 10.1201/9780367548445-20

Olaniyan, Olugbemi T., Adetunji, Charles O., Adeniyi, Mayowa J., & Hefft, Daniel Ingo (2022b). Computational intelligence in IoT healthcare. In *Deep learning, machine learning and iot in biomedical and health informatics* (1st ed.). First Published 2022. Imprint CRC Press. Pages 13. eBook ISBN 9780367548445. DOI: 10.1201/9780367548445-19

Orhan, K., Bayrakdar, I. S., Ezhov, M., Kravtsov, A., & Özyürek, T. (2020). Evaluation of artificial intelligence for detecting periapical pathosis on cone-beam computed tomography scans. *International Endodontic Journal*, *53*(5), 680–689. https://doi.org/10.1111/iej.13265

Ossowska, A., Kusiak, A., & Świetlik, D. (2022). Artificial intelligence in dentistry—Narrative review. *International Journal of Environmental Research and Public Health*, *19*(6), 3449. https://doi.org/10.3390/ijerph19063449

Oyedara, Omotayo Opemipo, Adeyemi, Folasade Muibat, Adetunji, Charles Oluwaseun, & Elufisan, Temidayo Oluyomi (2022). Repositioning antiviral drugs as a rapid and cost-effective approach to discover treatment against SARS-CoV-2 infection. In *Medical biotechnology, biopharmaceutics, forensic science and bioinformatics* (1st ed.). First Published 2022. Imprint CRC Press. Pages 12. eBook ISBN 9781003178903. DOI: 10.1201/9781003178903-10

Patcas, R., Bernini, D. A. J., Volokitin, A., Agustsson, E., Rothe, R., & Timofte, R. (2019). Applying artificial intelligence to assess the impact of orthognathic treatment on facial attractiveness and estimated age. *International Journal of Oral and Maxillofacial Surgery*, *48*(1), 77–83. https://doi.org/10.1016/j.ijom.2018.07.010

Rana, A., Singhal, N., & Badotra, S. (2021). A review paper on artificial intelligence. *ACADEMICIA: An International Multidisciplinary Research Journal*, *11*(10), 249–256.

Ramesh, A. N., Kambhampati, C., Monson, J. R. T., & Drew, P. J. (2004). Artificial intelligence in medicine. *Annals of the Royal College of Surgeons of England*, *86*(5), 334–338. https://doi.org/10.1308/147870804290

Seto, E., Leonard, K. J., Cafazzo, J. A., Barnsley, J., Masino, C., & Ross, H. J. (2012). Developing healthcare rule-based expert systems: Case study of a heart failure telemonitoring system. *International Journal of Medical Informatics*, *81*(8), 556–565. https://doi.org/10.1016/j.ijmedinf.2012.03.001

Setzer, F. C., Shi, K. J., Zhang, Z., Yan, H., Yoon, H., Mupparapu, M., & Li, J. (2020). Artificial intelligence for the computer-aided detection of periapical lesions in cone-beam computed tomographic images. *Journal of Endodontics*, *46*(7), 987–993. https://doi.org/10.1016/j.joen.2020.03.025

Sharma, S. (2019). Artificial intelligence in dentistry: The current concepts and a peek into the future. *International Journal of Contemporary Medical Research [IJCMR]*, *6*. https://doi.org/10.21276/ijcmr.2019.6.12.7

Sukegawa, S., Yoshii, K., Hara, T., Matsuyama, T., Yamashita, K., Nakano, K., Takabatake, K., Kawai, H., Nagatsuka, H., & Furuki, Y. (2021). Multi-task deep learning model for classification of dental implant brand and treatment stage using dental panoramic radiograph images. *Biomolecules*, *11*(6), 815. https://doi.org/10.3390/biom11060815

Xie, X., Wang, L., & Wang, A. (2010). Artificial neural network modeling for deciding if extractions are necessary prior to orthodontic treatment. *The Angle Orthodontist*, *80*(2), 262–266. https://doi.org/10.2319/111608-588.1

Yau, H. T., Tsou, L. S., & Tsai, M. J. (2006). Octree-based virtual dental training system with a haptic device. *Computer-Aided Design and Applications*, *3*(1–4), 415–424. https://doi.org/10.1080/16864360.2006.10738480

Yauney, G., Rana, A., Wong, L. C., Javia, P., Muftu, A., & Shah, P. (2019). Automated process incorporating machine learning segmentation and correlation of oral diseases with systemic health. *2019 41st Annual International Conference of the IEEE Engineering in Medicine and Biology Society (EMBC)*, 3387–3393. https://doi.org/10.1109/EMBC.2019.8857965

Yu, H. J., Cho, S. R., Kim, M. J., Kim, W. H., Kim, J. W., & Choi, J. (2020). Automated skeletal classification with lateral cephalometry based on artificial intelligence. *Journal of Dental Research*, *99*(3), 249–256. https://doi.org/10.1177/0022034520901715

8 Artificial Intelligence and Deep Learning Process and Drug Discovery

Olotu Titilayo
Adeleke University, Ede Osun State, Nigeria

Charles Oluwaseun Adetunji
Edo State University, Uzairue, Nigeria

8.1 INTRODUCTION

The limitation of biopharmaceutical industries has brought about novel technologies and inventions in chemical, biological, and mechanical engineering creating ease in the production of medications safe for humans; these technologies are achieved through artificial intelligence (AI) (Lo *et al.* 2020; (Adetunji *et al.* 2020; Oyedara *et al.* 2022; Adetunji *et al.* 2022a, 2022b, 2022c, 2022d, 2022e, 2022f, 2022g, 2022h; Olaniyan et al., 2022a and 2022b). A sequence read library contains next-gen sequencing data, National Cancer Institute Genomic Data Commission contains sequencing data on cancer and DriverML contains tools that point out cancer drivers. Data mining from PubMed identifies pointers from different disorders. Chemical databases like ChEMBL and DrugBank contain data on structures, biological and toxic properties, and their physicochemical properties including target interactions. AI with the aid of machine learning (ML) and deep learning (DL) provided the platform for computer-aided drug design software like CADD and Stanford HIV database. Protein-ligand interactions were predicted in a combination of a convolutional neural network and graph neural network and it was concluded that the obtained vector predicted vital amino acid residues for a potential drug candidate (Tsubaki *et al.* 2018).

DeepConv-DTI techniques now predict models of ligand-target complex with a combination of protein sequences and molecular fingerprints from datasets from KEGG, IUPHAR, and DrugBank. The use of artificial intelligence in the pharmaceutical industry is now seen as a powerful tool used in medicinal chemistry for drug discovery whose processes are previously long and complex. Artificial intelligence is used in drug discovery, refining pharmaceutical productivity, assisting in clinical traits, and aids the discovery of new drugs within a short time. The presence of highly efficient computing systems and algorithms, availability of biological and chemical data and insilco dug design software now create room for drug discovery that is fast, highly

DOI: 10.1201/9781003309468-8

efficient, and cost-effective, and the continuous innovations in artificial intelligence and deep learning system are greatly incorporated in various stages of drug discovery.

The transformation of machine learning to deep learning technologies creates the potential of managing big data with the use of AI and insilico drug discovery software leading to the newest drugs discoveries (Tripathi *et al.* 2021; Paul *et al.* 2021). Deep learning creates advanced potentials in drug discovery like ligand prediction, generation of viable chemical entities, and image analysis. Artificial intelligence like deep learning and modeling studies in big data models is the new solution for safety assessment of drug candidates which creates a clear understanding of vitro and in vivo clinical traits outcomes. The use of ML cuts across target validation, identification of exact biomarkers to methodologies, and methods arriving at precise predictions. Advancement in AI in big data of biological and chemical analysis provides a path for new drug creation and optimization which are significant to public health (Vamathevan *et al.* 2019; Zhu 2020).

Patel and Shah (2021) grouped the use of AI in the pharmaceutical industry into four categories, (a) for assessment of the severity of the diseases and determination of successful treatment of diseases, (b) for prevention and solving of complications during treatment, (c) as assistive technology to during treatment of patients, and (d) to ensure the use of particular instruments or chemical to increase safety during treatment. Artificial intelligence and big data are composed of deep learning and machine learning. Deep learning and machine learning create new methods and technologies that have emerged as a successful tool for advanced drug development. This chapter is an overview of advanced technologies in drug discovery and machine learning in drug discovery.

8.2 MACHINE LEARNING AND DEEP LEARNING IN DRUG DISCOVERY

Artificial intelligence involves simulation of human intelligence through insilico analysis, target identification, identification of ligands, and effectiveness of ligands in creating new drugs. Data derived from genetics, proteomics, and other life sciences allow the innovative use of Al in drug discovery (Mishra and Awasthi 2021). Analyzing big data is managed through Al and the deep learning process creating new insight into the use of these data (Solanki *et al.*, 2021). Big data are data that are huge and conventional data software is not able to analyze them due to their volume, variety, and velocity because it reproduces at a high rate (Gandomi and Haider 2015).

The application of Al in big data analytics with the aid of machine learning has made data analysis easier especially in the area of big biomedical data, extracting useful features, target identification, and gene expression for a better understanding of disease mechanisms (Lo *et al.* 2020). NCBI, Cancer Genome Atlas, and Arrayexpress are some of the repositors of big data containing genetic data (Gupta *et al.* 2021). Genome-wide association studies are used to determine genomic variants, help ascertain disease-associated genetic loci, and identify mutated genes as typical therapeutic targets. The Sequence Read library contains next-gen sequencing data, the National Cancer Institute Genomic Data Commission contains sequencing data on cancer, and DriverML contains tools that point out cancer drivers.

Data mining from PubMed identifies pointers from different disorders. A chemical database like ChEMBL and DrugBank contains data on structures, biological and toxic properties, and their physicochemical properties including target interactions. Also, there is protein data bank data of proteins, RNA, and DNA structure which are also used for protein-ligand interactions assessment (Gupta *et al.* 2021). Big data in pharmaceutical manufacturing create an opportunity for in-depth data mining through data extraction and collation, uniform formatting, and using various platform output interpreted that inform decisions about medicines to developing yielding efficient treatment. Innovations of AI into drug development in the biopharmaceutical industry reduce research cost and time of research span and assist in predicting possible potential side effects of these drugs. (Anthony 2016; Kshirsagar and Shah 2021).

AI embraces fuzzy systems, neural networks, or deep learning which are greatly applied in medicinal chemistry, and evolutionary computation. The drug discovery scenario isn't different from what was done in the last decade where a huge volume of data was generated in graphics processing computing; the huge volume of chemical and biological data with computerization of techniques sum up the computational intelligence or artificial intelligence (Lipinski *et al.* 2019). Computational methods allow medicinal chemistry and the use of molecular features to be related to the functionality of molecular targets and pharmacokinetics properties; the result obtained could assist in drug design and discovery (Maltarollo *et al.* 2017).

ML techniques are necessary for knowledge attainment of molecular properties of compounds and ML creates platforms for drug discovery procedures (Lima *et al.* 2016). Deep learning is related to layers of the network, the deeper the layer the deeper the network, and these networks are deep belief networks and convolutional neural networks, and multi-task learning (Lipinski *et al.* 2019).

Deep learning provides functional predictive models in drug discovery through managing large chemical libraries or data. Deep learning techniques screen modeling of potential drug candidates through virtual screening and quantitative structure-activity interactions (Zhang *et al.* 2016). In protein engineering, it's used to explore the function and structure of the protein, suggesting the structures within the molecules, and it predicts the function of raw data structures in electron density (Golkov *et al.* 2017). In sequencing technologies, huge genomic data are used for genomic modeling in precision medicine and drug design (Aliper *et al.* 2016).

They are used to predict medicinal chemistry endpoints, the interaction of protein-ligand, docking scoring, and screening. They predict toxicity parameters like solubility and specific toxicities in target properties with the aid of deep learning models (Capuzzi *et al.* 2016; Rifaioglu *et al.* 2018). AI with the aid of ML and DL provided the platform for computer-aided drug design software like CADD and the Stanford HIV database.

Machine learning entails deep learning of artificial neutral networks which are composed of arrays of processing structures called neurons. Deep learning neutral networks are made up of many hidden layers that indicate the depths of the networks and use deep specified architectures to learn higher-level data in an automated fashion. Deep learning allows ligand-based methodologies in drug development where a physicochemical property potential drug candidate is predicted and molecular descriptors are used to construct models that indicates the chemical biological

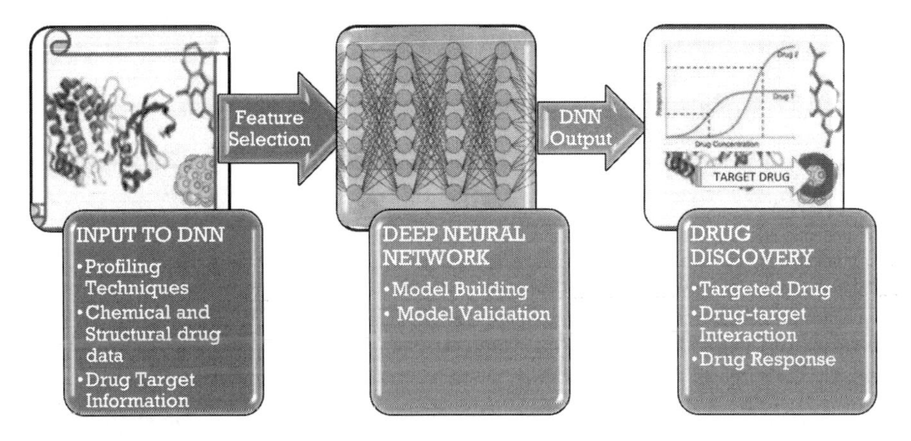

FIGURE 8.1 Deep neural network in deep learning applications to drug discovery.

structural activity. DBN models were used to predict drug-target interactions in the target and compounds using the dataset from DrugBank for repurposing drug studies (Wen *et al.* 2017). Data from used ChEMBL database were used to assay molecular targets in dataset curation and enzymes, neutral receptors, and other molecular targets were identified (Rifaioglu *et al.* 2018).

Protein-ligand interactions were predicted in a combination of convolutional neural network and graph neural network and it was concluded that the obtained vector predicted vital amino acid residues for a potential drug candidate (Tsubaki *et al.* 2018). DeepConv-DTI techniques by Lee *et al.* (2019) predicted models of ligand-target complex with a combination of protein sequences and molecular fingerprints from datasets from KEGG, IUPHAR, and DrugBank. AI removes bottlenecks in computational drug design enhancing the development of low-cost and more effective medication production. Moreover, ML tools are used to determine a three-dimensional protein target structure which is critical to the development of a novel drug like Google's DeepMind for the prediction of 3D structure from amino acid sequences (Smith *et al.* 2018; Jing *et al.* 2018). Quantum mechanics determines subatomic properties for estimating protein-ligand interactions (Kalaiarasi *et al.* 2019). In this present age, AI technologies have achieved great success in biomedical research, particularly in genomic analysis and drug design. Data mining, genomic analysis, target molecules predictions, designing new compounds. It saves time, is cost-effective, and treatments are appropriate and forecast like side effects in drug development (Figure 8.1).

REFERENCES

Adetunji, Charles Oluwaseun, Mitembo, William Peter, Egbuna, Chukwuebuka, Narasimha Rao, G.M. (2020). In silico modeling as a tool to predict and characterize plant toxicity. In Andrew G. Mtewa, Chukwuebuka Egbuna, & G.M. Narasimha Rao (Eds.), *Poisonous plants and phytochemicals in drug discovery*. Wiley Online Library DOI: 10.1002/9781119650034.ch14

Adetunji, Charles Oluwaseun, Nwankwo, Wilson, Olayinka, Akinola Samson, Olugbemi, Olaniyan Tope, Akram, Muhammad, Laila, Umme, Olugbenga, Michael Samuel, Oshinjo, Ayomide Michael, Adetunji, Juliana Bunmi, Okotie, Gloria E., & Esiobu, Nwadiuto (Diuto) (2022h). Machine learning and behaviour modification for COVID-19. In *Medical biotechnology, biopharmaceutics, forensic science and bioinformatics* (1st ed.). First Published 2022. Imprint CRC Press. Pages 17. eBook ISBN 9781003178903. DOI: 10.1201/9781003178903-17

Adetunji, Charles Oluwaseun, Nwankwo, Wilson, Olayinka, Akinola Samson, Olugbemi, Olaniyan Tope, Akram, Muhammad, Laila, Umme, Samuel, Michael Olugbenga, Oshinjo, Ayomide Michael, Adetunji, Juliana Bunmi, Okotie, Gloria E., & Esiobu, Nwadiuto (Diuto) (2022f). Computational intelligence techniques for combating COVID-19. In *Medical biotechnology, biopharmaceutics, forensic science and bioinformatics* (1st ed.). First Published 2022. Imprint CRC Press. Pages 12. eBook ISBN 9781003178903. DOI: 10.1201/9781003178903-16

Adetunji, Charles Oluwaseun, Olaniyan, Olugbemi Tope, Adeyomoye, Olorunsola, Dare, Ayobami, Adeniyi, Mayowa J, Alex, Enoch, Rebezov, Maksim, Garipova, Larisa, & Shariati, Mohammad Ali. (2022a). eHealth, mHealth, and telemedicine for COVID-19 pandemic. In S.K. Pani, S. Dash, W.P. dos Santos, S.A. Chan Bukhari, & F. Flammini (Eds.), *Assessing COVID-19 and other pandemics and epidemics using computational modelling and data analysis*. Springer. DOI: 10.1007/978-3-030-79753-9_10

Adetunji, Charles Oluwaseun, Olaniyan, Olugbemi Tope, Adeyomoye, Olorunsola, Dare, Ayobami, Adeniyi, Mayowa J., Alex, Enoch, Rebezov, Maksim, Isabekova, Olga, & Shariati, Mohammad Ali (2022c). Smart sensing for COVID-19 pandemic. In S.K. Pani, S. Dash, W.P. dos Santos, S.A. Chan Bukhari, & F. Flammini (Eds.), *Assessing COVID-19 and other pandemics and epidemics using computational modelling and data analysis*. Springer. DOI: 10.1007/978-3-030-79753-9_9

Adetunji, Charles Oluwaseun, Olaniyan, Olugbemi Tope, Adeyomoye, Olorunsola, Dare, Ayobami, Adeniyi, Mayowa J., Alex, Enoch, Rebezov, Maksim, Koriagina, Natalia, & Shariati, Mohammad Ali (2022e). Diverse techniques applied for effective diagnosis of COVID-19. In S.K. Pani, S. Dash, W.P. dos Santos, S.A. Chan Bukhari, & F. Flammini (Eds.), *Assessing COVID-19 and other pandemics and epidemics using computational modelling and data analysis*. Springer. DOI: 10.1007/978-3-030-79753-9_3

Adetunji, Charles Oluwaseun, Olaniyan, Olugbemi Tope, Olorunsola Adeyomoye, Ayobami Dare, Mayowa J. Adeniyi, Enoch Alex, Maksim Rebezov, Ekaterina Petukhova, & Shariati, Mohammad Ali (2022b). Machine learning approaches for COVID-19 pandemic. In S.K. Pani, S. Dash, W.P. dos Santos, S.A. Chan Bukhari, & F. Flammini (Eds.), *Assessing COVID-19 and other pandemics and epidemics using computational modelling and data analysis*. Springer. DOI: 10.1007/978-3-030-79753-9_8

Adetunji, Charles Oluwaseun, Olaniyan, Olugbemi Tope, Adeyomoye, Olorunsola, Dare, Ayobami, Adeniyi, Mayowa J., Alex, Enoch, Rebezov, Maksim, Petukhova, Ekaterina, & Shariati, Mohammad Ali (2022d). Internet of Health Things (IoHT) for COVID-19. In S.K. Pani, S. Dash, W.P. dos Santos, S.A. Chan Bukhari, & F. Flammini (Eds.), *Assessing COVID-19 and other pandemics and epidemics using computational modelling and data analysis*. Springer. DOI: 10.1007/978-3-030-79753-9_5

Adetunji, Charles Oluwaseun, Olugbemi, Olaniyan Tope, Akram, Muhammad, Laila, Umme, Samuel, Michael Olugbenga, Oshinjo, Ayomide Michael, Adetunji, Juliana Bunmi, Okotie, Gloria E., Esiobu, Nwadiuto (Diuto), Oyedara, Omotayo Opemipo, & Adeyemi, Folasade Muibat (2022g). Application of computational and bioinformatics techniques in drug repurposing for effective development of potential drug candidate for the management of COVID-19. In *Medical biotechnology, biopharmaceutics, forensic science and bioinformatics* (1st ed.). First Published 2022, Imprint CRC Press. Pages 14. eBook ISBN 9781003178903. DOI: 10.1201/9781003178903-15

Aliper, A., Plis, S., Artemov, A., Ulloa, A., Mamoshina, P., and Zhavoronkov, A. (2016). Deep learning applications for predicting pharmacological properties of drugs and drug repurposing using transcriptomic data. *Mol Pharm, 13*, 2524–2530. doi: 10.1021/acs.molpharmaceut.6b00248

Anthony. (2016). Big data in medicine: The upcoming artificial intelligence. *Prog Pediatr Cardiol, 43*(2016), 91–94. doi: 10.1016/j.ppedcard.2016.08.021

Capuzzi, S. J., Politi, R., Isayev, O., Farag, S., and Tropsha, A. (2016). QSAR modeling of Tox21 challenge stress response and nuclear receptor signaling toxicity assays. *Front Environ Sci*, 4, 3. doi: 10.3389/fenvs.2016.00003

Gandomi, A., & Haider, M. (2015). Beyond the hype: Big data concepts, methods, and analytics. *Int J Inf Manage*, 35, 137–144. doi: 10.1016/j.ijinfomgt.2014.10.007

Golkov, V., Skwark, M. J., Mirchev, A., Dikov, G., Geanes, A. R., Mendenhall, J., et al. (2017). 3D deep learning for biological function prediction from physical fields. *arXiv: 1704.04039*

Gupta, R., Srivastava, D., Sahu, M., Tiwari, S., Ambasta, R.K., & Kumar, P. (2021). Artificial intelligence to deep learning: Machine intelligence approach for drug discovery. *Mol Divers*, 25, 1315–1360. doi: 10.1007/s11030-021-10217-3

Jing, Y., Bian, Y., Hu, Z. et al (2018) Deep learning for drug design: An artificial intelligence paradigm for drug discovery in the big data era. *AAPS J*, 20(3), 58. doi: 10.1208/s12248-018-0210-0

Kalaiarasi, C., Manjula, S., & Kumaradhas, P. (2019). Combined quantum mechanics/molecular mechanics (QM/MM) methods to understand the charge density distribution of estrogens in the active site of estrogen receptors. *RSC Adv*. DOI: 10.1039/c9ra08607b

Kshirsagar, A., & Shah, M. (2021). Anatomized study of security solutions for multimedia: Deep learning-enabled authentication, cryptography and information hiding. *Adv Secur Solut Multimed*, 10.1088/978-0-7503-3735-9CH7

Lee, I., Keum, J., & Nam, H. (2019). DeepConv-DTI: Prediction of drug-target interactions via deep learning with convolution on protein sequences. *PLoS Comput Biol*, 15, e1007129. doi: 10.1371/journal.pcbi.1007129

Lima, A. N., Philot, E. A., Trossini, G. H. G., Scott, L. P. B., Maltarollo, V. G., & Honorio, K. M. (2016). Use of machine learning approaches for novel drug discovery. *Expert Opin Drug Discovery* 11, 225–239. doi: 10.1517/17460441.2016.1146250

Lipinski, C. F., Maltarollo, V. G., Oliveira, P. R., da Silva Alberico, B. F., & Honorio, K. M. (2019). Advances and perspectives in applying deep learning for drug design and discovery. *Front Robot AI, 6*, 2296–9144. https://www.frontiersin.org/article/10.3389/frobt.2019.00108

Lo, Y.-C., Ren, G., Honda, H. L., & Davis, K. (2020) Artificial intelligence-based drug design and discovery. In *Cheminformatics and its applications*. doi: 10.5772/intechopen.89012

Maltarollo, V. G., Kronenberger, T., Wrenger, C., & Honorio, K. M. (2017). Current trends in quantitative structure–activity relationship validation and applications on drug discovery. *Future Sci OA* 3, FSO214. doi: 10.4155/fsoa-2017-0052

Mishra, D., & Awasthi, H. (2021). Artificial intelligence: A new era in drug discovery. *Asian J Pharm Res Dev, 9*(5), 87–92. doi: 10.22270/ajprd.v9i5.995

Olaniyan, Olugbemi T., Adetunji, Charles O., Adeniyi, Mayowa J., & Hefft, Daniel Ingo (2022a). Machine learning techniques for high-performance Computing for IoT applications in healthcare. In *Deep learning, machine learning and iot in biomedical and health informatics* (1st ed.). First Published 2022. Imprint CRC Press. Pages 13. eBook ISBN 9780367548445. DOI: 10.1201/9780367548445-20

Olaniyan, Olugbemi T., Adetunji, Charles O., Adeniyi, Mayowa J., & Hefft, Daniel Ingo (2022b). Computational intelligence in IoT healthcare. In *Deep learning, machine learning and iot in biomedical and health informatics* (1st ed.). First Published 2022. Imprint CRC Press. Pages 13. eBook ISBN 9780367548445. DOI: 10.1201/9780367548445-19

Oyedara, Omotayo Opemipo, Adeyemi, Folasade Muibat, Adetunji, Charles Oluwaseun, & Elufisan, Temidayo Oluyomi (2022). Repositioning antiviral drugs as a rapid and cost-effective approach to discover treatment against SARS-CoV-2 infection. In *Medical biotechnology, biopharmaceutics, forensic science and bioinformatics* (1st ed.). First Published 2022. Imprint CRC Press. Pages 12. eBook ISBN 9781003178903. DOI: 10.1201/9781003178903-10

Patel, V., & Shah, M. (2021). *A comprehensive study on artificial intelligence and machine learning in drug discovery and drug development.* Intelligent Medicine, ISSN 2667-1026

Paul, Sanap D., Shenoy, G., Kalyane, S., Kalia, D. K., & Tekade, R. K. (2021). Artificial intelligence in drug discovery and development. *Drug Discov Today*, *26*(1), 80–93. doi: 10.1016/j.drudis.2020.10.010

Rifaioglu, A. S., Atalay, V., Martin, M. J., Cetin-Atalay, R., & Dogan, T. (2018). DEEPScreen: high performance drug-target interaction prediction with convolutional neural networks using 2-D structural compound representations. *bioRxiv* 491365. doi: 10.1101/491365

Smith, J. S., Roitberg, A. E., & Isayev, O. (2018) Transforming computational drug discovery with machine learning and AI. *ACS Med Chem Lett*, 9(11), 1065–1069

Solanki S, Dehalwar V, Choudhary J. (2021). Deep learning for spectrum sensing in cognitive radio. *Symmetry*, 13(1):147. https://doi.org/10.3390/sym13010147

Tripathi, N., Goshisht, M. K., Sahu, S. K., & Arora, C. (2021). Applications of artificial intelligence to drug design and discovery in the big data era: A comprehensive review. *Mol Divers*, 25(3), 1643–1664. doi: 10.1007/s11030-021-10237-z

Tsubaki, M., Tomii, K., & Sese, J. (2018). Compound–protein interaction prediction with end-to-end learning of neural networks for graphs and sequences. *Bioinformatics*, 35, 309–318. doi: 10.1093/bioinformatics/bty535

Vamathevan, J., Clark, D., Czodrowski, P. et al. (2019). Applications of machine learning in drug discovery and development. *Nat Rev Drug Discov*, *18*, 463–477.

Wen, M., Zhang, Z., Niu, S., Sha, H., Yang, R., Yun, Y., et al. (2017). Deep-learning-based drug-target interaction prediction. *J Proteome Res*, 16, 1401–1409. doi: 10.1021/acs.jproteome.6b00618

Zhang, R., Li, J., Lu, J., Hu, R., Yuan, Y., & Zhao, Z. (2016). Using deep learning for compound selectivity prediction. *Curr Comput Aided Drug Des* 12, 5. doi: 10.2174/1573409912666160219113250

Zhang, Y., Wei, Y., & Yang, Q. (2018). "Learning to multitask," in *Proceedings of the 32nd International Conference on Neural Information Processing Systems* (Montreal, QC), 5776–5787.

Zhu, H. (2020). Big data and artificial intelligence modeling for drug discovery. *Annu Rev Pharmacol Toxicol* 60, 573–589. doi: 10.1146/annurev-pharmtox-010919-023324

9 Clinical Trials in Health Informatics

Frank Abimbola Ogundolie
Department of Biotechnology, Baze University
Abuja, Nigeria

Ale Oluwabusolami
Federal Medical Centre, Jabi, Nigeria
University of North Carolina at Chapel Hill, USA

Akinmoju Olumide Damilola
Ladoke Akintola University of Technology Teaching Hospital, Ogbomosho, Nigeria
Monefiore St. Luke's Cornwall Hospital, Newyork, USA

Onotu Roseline Ohunene
Federal Medical Centre, Jabi, Nigeria

9.1 INTRODUCTION

Over time, the emergence of technology has led to its application also in the field of medicine. Technology is now being incorporated into several aspects of medicine such as electronic health records (EHRs), telemedicine, decision support systems, global health informatics, etc., to improve the delivery of healthcare. This marriage between technology and healthcare has been described as health informatics. Other names for health informatics include clinical/medical informatics and it is described as the interrelation between the healthcare practice and use of computers and information technology (IT) in general (Michigan Tech, 2022). It is a key part of medicine that deals with "the resources, technology, and methods required to enhance the acquisition, storage, retrieval, and utilization of data in health and biomedicine" (Michigan Tech, 2022).

Public health researchers are also one of the categories of people to benefit from this development. It is easier to conduct research with secondary data obtained through health informatics. Health informatics has also helped in the advancement of clinical trials. Clinical trials are a type of research study that is carried out to evaluate medical, surgical or behavioural intervention effectiveness (National Institute on Aging, 2020). They evaluate a new treatment and its effects or outcomes on human health. Hospitals now make use of EHRs and apply IT in the computation of patients'

DOI: 10.1201/9781003309468-9

data which provide easy access to information and use of information for health research purposes and public health.

Clinical trials as a type of research method are very cumbersome and very expensive. It is very costly and time-demanding if it has to be done on-site. Site-based cost accounts for 60–75% of typical clinical trial costs (Eisenstein et al., 2005). It involves finding subjects for the study, identifying them, verifying if they meet inclusion and exclusion criteria, and also following up on information to monitor effects or outcomes. This is why many clinical trials have now integrated secondary data use in conducting clinical trials, which helps to save time and costs.

However, despite the expansion of health informatics into clinical trials, not many authors have explicitly acknowledged its role in making clinical trials easier. This chapter aims to see how the invention of health informatics has been helpful in the conduction of clinical trials and to also highlight specific authors that utilized health informatics in their published works.

9.2 HEALTH INFORMATICS

Health informatics has recently been gaining recognition and some awareness as an important component of healthcare and healthcare delivery and for its potential to change the course of healthcare delivery in the nearest future. There is an increase in the volume of patient data and the need to efficiently coordinate these various data to help physicians put results together and forge better and more advanced ways to deliver quality healthcare.

Health informatics is indeed a diverse field that entails the gathering, processing, storage, and presentation of data, as well as its application with the aim of making health and healthcare delivery more effective and efficient. It's a branch of medicine focused on making the best use of data, typically with the help of information technology, to enhance personal health, universal healthcare, public health, and health research (Hersh, 2009). Health informatics may improve health through saving cost, improved management of information, patient satisfaction, better outcomes, and patient confidentiality.

It was initially introduced in the mid-20th century, in the aftermath of World War II. Several doctors and researchers were looking into ways in which diseases could be diagnosed with the aid of computers. Generally, informatics is the process of integration of people, data and technological systems to generate, organize, store and possible future retrieval of information.

There are enormous challenges currently facing healthcare delivery and health systems, to ensure that the health needs of the nation are reached, and healthcare remains easily accessible to everyone, irrespective of their ailments or background. From the increase in world population, which is growing and ageing to the presence of urbanization and development of antibiotic resistance, "which is poised" to bring a significant change to how healthcare is delivered, which could affect affordability and accessibility to healthcare. Without doubt, IT has been of benefit to many sectors, including the health sector (Kabakuş and Kara, 2016). According to research, the healthcare system invests about 2% of its income in IT, indicating that health informatics can still play a larger role with more funding of its cause.

In all this, the benefits that come with the use of IT in the health space, to support healthcare delivery and healthcare management and at the same time projecting these benefits to the masses while avoiding medical malpractices and ensuring safe, equitable, affordable healthcare delivery is becoming unwaveringly compulsory. The integration and adaptation of healthcare and technology with the aim of better handling health information and communication will undoubtedly improve productivity and efficiency and in turn better health outcomes.

The application of IT to health comes with numerous benefits which can be very easily overlooked. It is presumed that efficient use of computer technology in healthcare would inadvertently improve healthcare service delivery methods, and make healthcare more affordable and safer (Hovenga and Grain, 2016). Healthcare is now experiencing an overload of data and information, with excess data being collected and stored during the critical care process. Clinical information systems (CIS) that provide comprehensive medical documentation at the bedside may now gather and preserve a lot of this data.

At the level of the hospital, there is also an overload of information; new and updated policies at the administrative level have become overwhelming, which has increased complexity.

Various clinical studies are ongoing while thousands of data are being released monthly and yearly. There are various clinical trials and literature to keep up with. Human beings are still prone to errors; however, health informatics is here to solve this overload (Imhoff, 2002).

9.3 IMPORTANCE OF HEALTH INFORMATICS IN MEDICINE

It is evident that the flourishing field of health informatics has experienced significant growth since its origins in the 1950s, highlighted by Masic's 2020 reference to the use of computer and information processing systems in medicine. Who would have thought that one day, a specialty would bridge the gap between medicine, computer science and data analysis? A field so pivotal in the revolutionization of modern medicine, is the cross section of artificial intelligence and health informatics. Health informatics is transforming how patient data is managed, leading to improved outcomes, enhanced communication among providers, and more personalized care. One could say that there is a symbiotic relationship between the relatively new field of health informatics and medicine as a whole.

The relevance of health informatics to medicine is multitudinous; however, it will be encapsulated in this chapter by discussing three vital ways by which the medical field is influenced by health informatics. The first is a reduction in total cost. By automating labour-intensive or time-consuming procedures, health informatics speeds up processes by offering services concurrently and reduces the amount of human effort necessary (Kabakuş and Kara, 2016). It is worthy to note that prompt rendering of health services from all fronts is usually delayed by the asynchronous delivery of healthcare by all parties involved. A typical healthcare institution relying on the traditional system of record-keeping such as patient folder requires healthcare to be delivered after orders are charted. The uncertainties in the time spent charting and communicating this vital information from one party to the other may lead to unnecessary time delays and wastage of personnel. In 2005, a systematic evaluation of relevant research found that documenting on a computer reduced documentation time significantly (weighted average, 234.0% working shifts) (Poissant et al., 2005).

The intervention of health informatics in healthcare cost is not only beneficial to the healthcare providers but also to patients. The growth of patient-centred care, a concept that is now widely accepted as the ideal standard of healthcare across borders, has been bolstered by the incorporation of medical informatics through patient privacy and patient empowerment. Patients' electronic medical records can be made highly secured, only accessible to a few healthcare providers. This boosts patients' confidence in the system, increases accountability and reduces the frequency of medico-legal disputes. Through the advent of bioinformatics, patients can be carried along in the entire process of healthcare.

Interoperability, a concept that allows both healthcare providers and patients to access medical records plays a vital role in the evolvement of patient-centred care. Through the accessibility of mobile devices and applications, patients can now interact with their healthcare providers, and access information concerning their health and management. This also allows physicians to have access to vital home records like ECG, weight, blood sugar, oxygen saturation, etc. which reduces the frequency of visits, enhances the effectiveness of care, and allows early intervention in emergency cases.

Increased effectiveness in healthcare delivery is one way health informatics has impacted the field of medicine. The incorporation of bioinformatics into medicine has led to improved outcomes, a reduction in medical errors, and improved coordination among healthcare workers (Pharmapproach, 2020) thus leading to prompt and timely diagnosis and clinical intervention, improving the healthcare industry at large.

9.4 ROLE OF HEALTH INFORMATICS IN CLINICAL TRIALS

Health informatics is the branch of medicine concerned with the utilization of IT to enhance the delivery of healthcare. Its role in improving healthcare delivery has been established through its application in public health, decision-making, problem-solving, biomedical care, and clinical trials inclusive (Kulikowski et al., 2012). Many clinical research trials are not focused on therapy only but also on the natural course of illness, diagnostic criteria, the importance of patient education, and long-term surveillance. Such investigations/research need clinical records as well as data held in numerous administrative systems for patient care. (Coorevits et al., 2013). With health informatics, there is less chance of missing information required for clinical trials.

Health informatics has also played a vital role in study participants' recruitment for clinical trials. Meeting the goal number of participants required for research has been an issue for many clinical studies. Campbell et al. from their review of over 100 trials found that less than 30% of authors achieved the targeted number of participants for the study (Campbell et al., 2007). This drawback has necessitated consideration of other options for recruitment in which health informatics seems to provide a better alternative way. With health informatics, patients' data and information are already existing and readily available.

Authors have been able to create a command algorithm to filter out patients that are eligible for clinical trials. Carlson et al. (1995) did this in their study among HIV/AIDS patients and Ohno-Machado et al. (1999) in their study among breast cancer patients. Another exciting feature is the use of alarms that can be incorporated in apps to alert authors when a patient becomes eligible to fit into a study, an example of this is the sepsis-alerting system(Thompson et al., 2003). A model of a digital medical record-based clinical research alerting system that aids in clinical trial recruiting has now been adopted and deployed by several institutions. (Dugas et al., 2008; Embi et al. 2005a). The benefits of an increase in recruitment rates due to health informatics have been documented by previous authors (Embi et al., 2005b). All of these have highlighted the ease of patient recruitment for clinical trials research.

Additionally, patient retention is another challenge that health informatics has helped in tackling in clinical trials through the reduction in loss of participants to follow up. This loss is very significant as the average recruitment dropout rate for clinical trials has been reported to be 30% (Alexander, 2013; Neumann, 2014). Losing participants has a significant impact on a study's noteworthiness, as well as less clear results concerning the trial's goals and objectives. The geographical barrier is one of the major factors responsible for this loss that health informatics has helped to tackle in helping to increase recruitment rates. Patients will usually be reluctant in travelling down to a trial centre especially those from low socioeconomic class; however, with the use of electronic records, this can be taken care of. Health informatics helps to reduce the burden and cumbersomeness that may frustrate study participants, giving them less reason to drop out as they can input required information right from their comfort zone (Geng et al., 2010).

Similarly, health informatics has helped in the significant reduction of the time required for the conduct of a clinical trial. The long length of time has been a drawback of clinical trials as it takes a longer time to get it completely done but with health informatics, it is faster than usual as recruitment time is shorter. It helps to save time wasted in waiting to get adequate participants for a study to commence. The need to operate enrolment sites for a long time is taken care of, which also helps to reduce human resources and costs significantly. The present state of clinical studies highlights the importance of speedy solutions to reduce the cost and duration of typical studies. Health information systems, telehealth services, and digital health applications, among other newly created health informatics approaches, have recently been employed in many clinical trials to optimize various aspects of the trial process.

9.5 CHALLENGES FACING HEALTH INFORMATICS IN CLINICAL TRIALS

Advancements in health informatics technology over the last few decades have unquestionably changed healthcare delivery and how health data is stored. Nowadays, medical procedures allow for the sharing and preservation of a large volume of patient-specific data in electronic health records or auxiliary databases, including data from digital imaging tests and genetic sequence data. These data are available for biomedical research transformation and are not only for healthcare. However, because most clinical fundamental research data are shown in different systems, clinicians and researchers typically experience some difficulty obtaining and presenting information. (Prokosch & Ganslandt, 2009). Poor workflow management in the clinical research labs has also posed challenges to clinical research, decision-making and result evaluation. Health informatics' contribution to scientific research would be difficult in the absence of trustworthy and efficient reuse of electronic health records.

Another challenge that is causing drawbacks to the utilization of health informatics in clinical trials is the uncertainty about regulatory requirements and institutional policies. Institutional policies make provision for concerns about the protection of privacy of patients' data which affects the accessibility of data for clinical trial use. Regulations are also made for the security of patients' data which can limit accessibility. Another challenge of health informatics is the need for unified electronic health records that attends to the need for clinical research and clinical care. Oftentimes, during data exchange, alterations can be made to patients' data by operators who have access to the records. These alterations can cause a loss of data integrity leading to errors. There is a need for precision and accuracy in clinical trials, hence, its limited use.

9.6 SPECIFIC AUTHORS THAT HAVE USED HEALTH INFORMATICS IN CLINICAL TRIALS

Several authors have started integrating health informatics into their clinical research and some of the work has been highlighted below.

Geva et al. (2020) set out to integrate a high sensitivity natural language processing pipeline to detect potential adverse effects of drugs by identifying mentions of medications, signs and symptoms potentially indicative of adverse drug effects (ADEs) in patients' clinical notes. This system was referred to as the adverse drug event presentation and tracking (ADEPT) system. They found that the time taken to review potential adverse drug effects was shortened as it took a rotation of 4 minutes per patient compared to the traditional time of 15 to 23 minutes as reported studies by Ledieu et al. (2018) and Thevelin et al. (2018) thereby increasing the efficiency of pharmacovigilance. The ADEPT is a two-step process that allows for greater sensitivity and maintaining adequate specificity. However, this study was limited by the inability to directly compare a manual review of charts to potential adverse drug effects as there is a possible misrepresentation of a symptom that may be related to a patient's disease which can be misclassified as adverse drug effect.

Cho et al. (2020) also utilized a form of health informatics in their clinical trial research by examining the usability of consumer health informatics following the completion of a clinical trial. They tested the efficiency of a mobile health app in improving health outcomes among HIV associated non-AIDS conditions on how to self-manage symptoms through video information providers. The control group and intervention group were assessed on perceived usefulness and perceived ease of use of the mobile app; however, the control group was not shown self-care programs. Both groups perceived the app as being easy to use and useful. Both also found it useful in monitoring symptoms and enhancing communication with the clinical providers. However, only the intervention group perceived the app to be useful for improving overall health and long-term symptoms management. The findings showed how participants were able to integrate the app into their daily routine activities and use the app to support, track, monitor and self-manage symptoms related to HIV non-AIDS conditions.

Beck et al. (2017) also conducted a study that involved continuous glucose monitoring versus usual care in patients with type 2 diabetes mellitus receiving multiple daily insulin injections. Continuous glucose monitoring is a cutting-edge technology that continuously monitors a patient's glucose level using a sensor placed into subcutaneous fat, which records glucose every five minutes and communicates the data to a receiver that can build a graph based on the results. The control group monitored their glucose just four times daily. This typifies health informatics applications. This study was conducted because a previous study was done for type 1 diabetes mellitus which showed a significant reduction in hypoglycaemia, hyperglycaemia, HBA1C and glucose variability in those using it. Results for this study however showed a reduction in HBA1C for type 2 DM but no significant difference in hypoglycaemia or quality of life outcomes. This trial was able to be conducted in 25 endocrinology centres in North America because of the ease of transmission of data.

Lyons et al. (2015) also identified the shortfall of clinical assessment tools applied at visits for patients with dementia which then necessitated the studying of the efficiency of pervasive computing technologies to continuously assess Alzheimer's disease progression. It is to track activities and behaviours and measure relevant intra-individual changes. Sensors were installed in almost 500 homes over four years and monitored gait and mobility, sleep and activity, and medication adherence. Outcomes were mood, cognitive function, and loneliness. The goal is to see if this intervention can help to detect early or prodromal cognitive decline in older adults living in a typical community setting and how the best technology can be used. This allows capturing a wealth of data about patient status and daily activities of patients which were previously not possible.

Varnfield et al. (2014) also conducted a randomized control trial using a smartphone-based home care model to examine cardiac rehabilitation in post-myocardial infarction patients against the traditional centre-based rehabilitation program. Patients were monitored for six weeks and six months. Rehabilitation programs involved exercise monitoring, motivational deliveries weekly mentoring consultations. Other outcomes compared were modifiable lifestyle factors, biomedical risk factors, and health-related quality of life. Care Assessment Platform of Cardiac Rehabilitation (CAP-CR) which is the smartphone-based home care model had significant high uptake, adherence, and completion rates compared eith the traditional care rehabilitation. CAP-CR was also as effective in improving physiological and psychological health outcomes as traditional care rehabilitation which further reinforces CAP-CR as a viable option for optimizing the use of cardiac rehabilitation services.

9.7 HEALTH INFORMATICS TODAY

Health informatics has proved to be an improvement in modern medicine since its invention. Investing in health informatics, though capital intensive, is capable of saving the medical sector so much more in terms of cost, time and energy. Health informatics has proved to be beneficial in diverse ways and more, ranging from its contribution to healthcare delivery by improving patient-centred care and also furthering the effective delivery of healthcare. It is easy to say that clinical trials are the bedrock of science, without the knowledge and skills to perform clinical trials there would certainly be no growth in any aspect of science, from technology to medicine and even outside the scope of science itself.

Every new intervention in the practice of medicine and the discovery of new treatments required or started with people volunteering for clinical trials. A lot of research is still going on in the world of science, ways to further improve our day to day lives, and health, and easier ways to carry out various activities are all in the process of discovery. According to one of the great scientists to ever live, Albert Einstein, "the important thing is to never stop questioning". Everyday questions arise and discoveries and studies need to be made to further improve the world in which we live. Health informatics has been of great benefit and it is still a work in progress, large volumes of information are being garnered every day from various research work being done, and we continue to require the role health informatics plays to store this large volume of information while making it easily accessible whenever needed.

With more information being adequately stored, and being easily accessible, clinical trials become easier with more relevant data that can be easily referenced in the future without having to source new data or new participants to acquire such data from. This in turn would significantly shorten the time spent on performing clinical trials. As technology grows, so do needs for instant solutions to problems. The recent COVID-19 pandemic is a vivid example which took the world unawares and shook the healthcare delivery system all over the world, and the need for an instant solution was needed.

Under normal circumstances, it takes 10 to 15 years to build an effective and acceptable vaccine; however, this was impossible due to how the aforementioned virus was ravaging the health system of the global community, and this led to a vaccine being developed in under a year. This is no small feat but was, however, made possible through the availability of enough resources, including financial resources and more importantly infrastructures put in place to further aid clinical trials. Although medical research still has a long way to go, and beyond the field of medicine, the impact of health informatics is significant now and will be even more crucial in the future.

9.8 CONCLUSION

In this chapter, we looked into health informatics and its significance in the improvement of classical medicine. Drug design, discovery, and development have today been improved due to the technological advancements and provision/storage and analysis of health data. Though more is still expected in this regard because the dataset produced can be gender, tribe, and race biased and this will surely affect clinical trials and hence the outcome of clinical studies.

REFERENCES

Alexander, W. (2013). The uphill path to successful clinical trials: Keeping patients enrolled. *P & T: A Peer-Reviewed Journal for Formulary Management*, *38*(4), 225–227.

Beck, R. W., Riddlesworth, T. D., Ruedy, K., Ahmann, A., Haller, S., Kruger, D., McGill, J. B., Polonsky, W., Price, D., Aronoff, S., Aronson, R., Toschi, E., Kollman, C., & Bergenstal, R. (2017). Continuous glucose monitoring versus usual care in patients with type 2 diabetes receiving multiple daily insulin injections. *Annals of Internal Medicine*, *167*(6), 365–374. https://doi.org/10.7326/M16-2855

Campbell, M. K., Snowdon, C., Francis, D., Elbourne, D., McDonald, A. M., Knight, R., Entwistle, V., Garcia, J., Roberts, I., Grant, A., Grant, A., & STEPS Group. (2007). Recruitment to randomised trials: Strategies for trial enrollment and participation study. The STEPS study. *Health technology assessment (Winchester, England)*, *11*(48), iii, ix–105. https://doi.org/10.3310/hta11480

Carlson, R. W., Tu, S. W., Lane, N. M., Lai, T. L., Kemper, C. A., Musen, M. A., & Shortliffe, E. H. (1995). Computer-based screening of patients with HIV/AIDS for clinical-trial eligibility. *The Online Journal of Current Clinical Trials*, *4*, 179. https://pubmed.ncbi.nlm.nih.gov/7719564/

Cho, H., Porras, T., Flynn, G., & Schnall, R. (2020). Usability of a consumer health informatics tool following completion of a clinical trial: Focus group study. *Journal of Medical Internet Research*, *22*(6), e17708. https://doi.org/10.2196/17708

Coorevits, P., Sundgren, M., Klein, G. O., Bahr, A., Claerhout, B., Daniel, C., Dugas, M., Dupont, D., Schmidt, A., Singleton, P., Moor, G. D., & Kalra, D. (2013). Electronic health records: New opportunities for clinical research. *Journal of Internal Medicine*, *274*(6), 547–560. https://doi.org/10.1111/joim.12119

Dugas, M., Lange, M., Berdel, W. E., & Müller-Tidow, C. (2008). Workflow to improve patient recruitment for clinical trials within hospital information systems – A case study. *Trials*, *9*, 2. https://doi.org/10.1186/1745-6215-9-2

Eisenstein, E. L., Lemons, P. W., Tardiff, B. E., Schulman, K. A., Jolly, M. K., & Califf, R. M. (2005). Reducing the costs of phase III cardiovascular clinical trials. *American Heart Journal*, *149*(3), 482–488. https://doi.org/10.1016/j.ahj.2004.04.049

Embi, P. J., Jain, A., Clark, J., Bizjack, S., Hornung, R., & Harris, C. M. (2005a). Effect of a clinical trial alert system on physician participation in trial recruitment. *Archives of Internal Medicine*, *165*(19), 2272–2277. https://doi.org/10.1001/archinte.165.19.2272

Embi, P. J., Jain, A., Clark, J., & Harris, C. M. (2005b). Development of an electronic health record-based clinical trial alert system to enhance recruitment at the point of care. *AMIA Annual Symposium Proceedings*, 231–235.

Geng, E. H., Bangsberg, D. R., Musinguzi, N., Emenyonu, N., Bwana, M. B., Yiannoutsos, C. T., Glidden, D. V., Deeks, S. G., & Martin, J. N. (2010). Understanding reasons for and outcomes of patients lost to follow-up in antiretroviral therapy programs in Africa through a sampling-based approach. *Journal of Acquired Immune Deficiency Syndromes (1999)*, *53*(3), 405–411. https://doi.org/10.1097/QAI.0b013e3181b843f0

Geva, A., Stedman, J. P., Manzi, S. F., Lin, C., Savova, G. K., Avillach, P., & Mandl, K. D. (2020). Adverse drug event presentation and tracking (ADEPT): Semiautomated, high throughput pharmacovigilance using real-world data. *JAMIA Open*, *3*(3), 413–421. https://doi.org/10.1093/jamiaopen/ooaa031

Hersh, W. (2009). A stimulus to define informatics and health information technology. *BMC Medical Informatics and Decision Making*, *9*, 1–6.

Hovenga, E., & Grain, H. (2016). Learning, training and teaching of health Informatics and its evidence for informaticians and clinical practice. *Evidence-Based Health Informatics*, *222*, 336–354.

Imhoff, M. (2002). Health informatics. In: Sibbald, W.J., Bion, J.F. (eds) *Evaluating Critical Care: Update in Intensive Care Medicine*, vol 35. Springer, Berlin, Heidelberg. https://doi.org/10.1007/978-3-642-56719-3_18

Kabakuş, A. T., & Kara, R. (2016). The importance of informatics for the health care industry. *Acta Medica Anatolia*, *4*(3), 124–125. https://doi.org/10.5505/actamedica.2016.81300

Kulikowski, C. A., Shortliffe, E. H., Currie, L. M., Elkin, P. L., Hunter, L. E., Johnson, T. R., Kalet, I. J., Lenert, L. A., Musen, M. A., Ozbolt, J. G., Smith, J. W., Tarczy-Hornoch, P. Z., & Williamson, J. J. (2012). AMIA Board white paper: Definition of biomedical informatics and specification of core competencies for graduate education in the discipline. *Journal of the American Medical Informatics Association*, *19*(6), 931–938. https://doi.org/10.1136/amiajnl-2012-001053

Ledieu, T., Bouzillé, G., Thiessard, F., Berquet, K., Van Hille, P., Renault, E., Polard, E., & Cuggia, M. (2018). Timeline representation of clinical data: Usability and added value for pharmacovigilance. *BMC Medical Informatics and Decision Making*, *18*(1), 86. https://doi.org/10.1186/s12911-018-0667-x

Lyons, B. E., Austin, D., Seelye, A., Petersen, J., Yeargers, J., Riley, T., Sharma, N., Mattek, N., Wild, K., Dodge, H., & Kaye, J. A. (2015). Pervasive computing technologies to continuously assess alzheimer's disease progression and intervention efficacy. *Frontiers in Aging Neuroscience*, *7*, 102. https://doi.org/10.3389/fnagi.2015.00102

Masic, I. (2020). The history of medical informatics development - an overview. *International Journal on Biomedicine and Healthcare*, *8*(1), 37. https://doi.org/10.5455/ijbh.2020.8.37-52

Michigan Tech. (2022). *What is health informatics?* Michigan Technological University. https://www.mtu.edu/health-informatics/what-is/

National Institute on Aging. (2020). *What are clinical trials and studies?* National Institute on Aging. https://www.nia.nih.gov/health/what-are-clinical-trials-and-studies

Neumann, U. (2014). *Patient-centred clinical trials*. Campbell Pharmaceutical Meeting Series, Rutgers Business School. https://www.business.rutgers.edu/sites/default/files/documents/lerner/campbell-seminar-presentation-ulrich-neumann.pdf

Ohno-Machado, L., Wang, S. J., Mar, P., & Boxwala, A. A. (1999). Decision support for clinical trial eligibility determination in breast cancer. *Proceedings of the AMIA Symposium*, 340–344.

Pharmapproach. (2020, June 2). 6 benefits of health informatics. *Pharmapproach.Com*. https://www.pharmapproach.com/6-benefits-of-health-informatics/

Poissant, L., Pereira, J., Tamblyn, R., & Kawasumi, Y. (2005). The impact of electronic health records on time efficiency of physicians and nurses: a systematic review. *Journal of the American Medical Informatics Association: JAMIA*, *12*(5), 505–516. https://doi.org/10.1197/jamia.M1700

Prokosch, H. U., & Ganslandt, T. (2009). Perspectives for medical informatics. Reusing the electronic medical record for clinical research. *Methods of Information in Medicine*, *48*(1), 38–44.

Thevelin, S., Spinewine, A., Beuscart, J., Boland, B., Marien, S., Vaillant, F., Wilting, I., Vondeling, A., Floriani, C., Schneider, C., Donzé, J., Rodondi, N., Cullinan, S., O'Mahony, D., & Dalleur, O. (2018). Development of a standardized chart review method to identify drug-related hospital admissions in older people. *British Journal of Clinical Pharmacology*, *84*(11), 2600–2614. https://doi.org/10.1111/bcp.13716

Thompson, D. S., Oberteuffer, R., & Dorman, T. (2003). Sepsis alert and diagnostic system: Integrating clinical systems to enhance study coordinator efficiency. *Computers, Informatics, Nursing: CIN*, *21*(1), 22–26; quiz 27–28. https://doi.org/10.1097/00024665-200301000-00009

Varnfield, M., Karunanithi, M., Lee, C.-K., Honeyman, E., Arnold, D., Ding, H., Smith, C., & Walters, D. L. (2014). Smartphone-based home care model improved use of cardiac rehabilitation in postmyocardial infarction patients: Results from a randomised controlled trial. *Heart*, *100*(22), 1770–1779. https://doi.org/10.1136/heartjnl-2014-305783

10 SmartCare and Its Advantages in Biomedical Technologies and Health Informatics

Nyejirime Young Wike
State University of Medical and Applied Sciences, Igbo-eno, Enugu

Charles Oluwaseun Adetunji
Edo State University, Uzairue, Nigeria

Olugbemi T. Olaniyan
Rhema University, Aba, Nigeria

Juliana Bunmi Adetunji
Osun State University, Osogbo, Nigeria

Oluwafemi Adebayo Oyewole
Federal University of Technology, Minna, Nigeria

K. I. T. Eniola
Joseph Ayo Babalola University, Ikeji Arakeji, Nigeria

10.1 INTRODUCTION

As a result of the massive rise in the number of people, conventional healthcare cannot meet people's demands. Hospital facilities are not accessible or inexpensive to all, irrespective of the fact that there are very good facilities and innovative technologies. Technology mainly aims to assist consumers by informing people regarding their health situation and making them aware of their well-being. Individuals with SmartCare can personally handle various emergency conditions (Mohanty et al. 2016). It focuses on increasing the patient's standard of living and satisfaction. SmartCare enables the most efficient use of obtainable resources. It allows for distant diagnosis as well as lowers the healthcare expenditure for the patient. It also

DOI: 10.1201/9781003309468-10

enables doctors to expand their professional skills across territorial limitations. With the emerging popularity of pervasive computing, a functional smart health service ensures that its residents live a healthy life.

Software, activities, and interpretation design are the three major categories in which SmartCare features are defined. Software designs must enable stable operation among mobile applications and detectors, create a tailored connection between detectors and the recipient's personal computer, and protect data. Activities must be app-adaptable, with genuine surveillance, appropriate distribution, hypersensitivity, increased productivity with less energy consumption, and a smart factory. To enhance the customer encounter, interpretation design facilities must be able to build personality characteristics derived from previously obtained data, handle speech recognition methods, and possess widespread computational power (Bader et al. 2016; Banerjee and Gupta 2015).

Employing mobile broadband connections in multiple applications leads to effective structures for smart healthcare (Zhu et al. 2016). The systems and methods used to implement intelligent medical coverage have different aspects. Arrangement, structure, and context are the three important study aspects in medical systems investigation. The assemblage of numerous physiological parts in maximum compatibility, which can be utilized to tackle important challenges, is referred to as healthcare setup. Multiple processor arrays could be designed to leverage these setups in smooth medical data centers by placing the relevant sensing devices in the vicinity (Zhang et al. 2012). The tools and contexts wherein the healthcare structure is utilized make up a framework for intelligent hospital design. Network systems, computation systems, and cloud systems are all types of medical systems. The communication modules usually connect multiple designs and are referred to as "network systems." Computing systems can differ greatly depending on the techniques utilized.

The modern period is one of digitization. Conventional healthcare, which has biotechnology at its foundation, has started to digitalize and informationize as science and innovative concepts have advanced. SmartCare too has evolved, embracing a new era of digital technologies. SmartCare is more than just an improved technology; it is a number of co-transformations. According to Liu et al. (2018), medicalized modifications (from illness to caregiver treatment), informatization building adjustments (from diagnostic to local health informatization), adjustments in clinical treatment (from public to individualized monitoring), and modifications in the diagnostic and intervention theory too are examples of this transformation (beginning with concentrating on diagnosis to concentrating on protective medicine). Such developments concentrate on meeting patients' specific requirements while enhancing healthcare effectiveness, considerably enhancing the quality of medical satisfaction as well as representing contemporary drugs' advanced growth orientation.

The notion of a "Digital World," presented by IBM (Armonk, NY, USA) in 2009, gave birth to smart healthcare. Briefly described, Digital World is a smart network that detects data utilizing detectors, distributes it via the sensor networks, and analyzes it with powerful computers and cloud-based services (Martin et al. 2010). It can integrate social technologies to achieve a flexible and sophisticated understanding of organizational civilization. SmartCare is a health service program that utilizes wearable technology, sensor networks, and wireless broadband to flexibly decrypt messages, link individuals, equipment, and organizations, and then intelligently handle and react to healthcare environment requirements. SmartCare can foster communication among all stakeholders in the healthcare sector, ensuring that users access the benefits they require, assisting organizations in thinking critically and facilitating service provision. In a nutshell, smart healthcare is a greater level of health record architecture (Gong et al. 2013).

Several parties are involved in SmartCare, including patients and clinicians, clinics, and academic centers. This is a multi-dimensional natural being which includes illness preventive and surveillance, diagnostic, healthcare administrators, medical judgement calls, and clinical science. SmartCare is built on the foundation of data based on the web of Things, cellular technologies, cloud-based services, large data, fifth generation wireless (5G), microsystems, and cognitive computing, as well as current biotechnology. These innovations are frequently employed in SmartCare across all areas. Clients may use wearable devices to track their progress at all times, consult a physician via automated systems, such as using remote houses to run cloud servers, and physicians may provide a variety of sophisticated clinical information techniques to support and improve treatment.

Healthcare data may be managed by clinicians using an interconnected data system consisting of a lab data control method, picture preservation and communication networks, digital healthcare documentation, and other tools. Robotic surgery and 3D virtual techniques can help with better accuracy in surgery. According to the point of view of health facilities, broadcast recognition technology could be used to handle manpower substances and the distribution network, utilizing connected control platforms to gather data and aid decision-making. Users' perceptions could be improved by using telemedicine systems. From the standpoint of science-based research institutions, it is possible to employ data mining algorithms rather than hand pharmaceutical analysis, including using large data sets to discover acceptable participants (Pan 2019).

Health informatics refers to the analysis and implementation of technology to aid in the collection, storing, analysis, and use of data in order to enhance the methods or quality of healthcare operations (Hammond 2001). Even though the earliest application was most likely in the former Soviet Union, the phrase informatics is a rare example of backward French words, in which the English language usurped the word l'informatique from our French-speaking friends (Greenes and Shortliffe 1990). According to Detmar (2000), many of the most significant breakthroughs in health informatics could be viewed as requirements for providing effective patient-centered service, and that is the aim of many of these healthcare organizations. Digital data retrieval tools have made it easier for doctors to obtain the information they require (Tierney 2001). To make such information more valuable, it could be linked using a variety of predictive methods (Evans et al. 1995). After that, unprocessed or improved information can be gathered in a number of ways to make its significance more evident. Lastly, data can be made more valuable by presenting it in the right forms and at the right times.

10.2 ADVANTAGES OF SMARTCARE IN BIOMEDICAL TECHNOLOGIES AND HEALTH INFORMATICS

Data systems in the field of healthcare have advanced rapidly in the last ten years (Australian Institute of Health and Welfare [AIHW] 2012; Miller and Sim 2004). Such innovations, according to healthcare practitioners, would effectively enhance the gathering and distribution of medical data, including the use of medical data over book methods, allowing physicians to implement scientific proof of care (Hornbrook 2010; Lesk 2013). Lesk (2013) points out that nations like Denmark and the Netherlands, which have had complete electronic medical record service for nearly ten years, have documented advantages. Doctors have direct access to internet knowledge links, including assistance on therapeutic approaches, pharmaceutical vocabulary, code meanings, and ehealth publications, due to the use of this type of record rather than traditional data systems (Hornbrook 2010). According to Lesk (2013), Denmark has the lowest rate of medication mistakes compared to other Western nations that do not utilize electronic medical records since its registry informs doctors on risky substance interactions and responses depending on medical data stored in the database. Furthermore, it is critical for biomedical research since it facilitates entry to and evaluation of medical information stored in a single repository.

This SmartCare EMR initiative was employed in Zambia's HIV program, and it proved to have a number of benefits. It included: facilitating faster access to clients' data, which saves doctors time; facilitating the exchange of health information and HIV data via inclusive national files and revised client smart cards; as well as facilitating the control and assessment of HIV initiatives via the existence of global, regional, and municipal files (CDC 2010; Neame 2013). Further benefits, such as reduced bills due to little paperwork and the removal of several inquiries, since medical practitioners can swiftly sift and pick important documents to make timely judgments, mean SmartCare has made information utilization faster (Hornbrook 2010).

Another benefit of employing SmartCare is that it has become possible to gain the number of clients scheduled for assessment by just running a system review, as compared to the use of a sheet approach, where caregivers had to painstakingly come up with a number from the incident record (WHO 2013). A definitive breakdown of clients scheduled for assessment also assists in identifying and following up on others who miss scheduled visits, lowering the number of people who stop taking their medications and lowering the risk of infectious diseases (MOH 2012). SmartCare has made it simpler to analyze a whole batch of clients at a clinic rather than selecting, as is the situation with sheet files in just about all circumstances, because analyzing all casefiles in a specified timeframe is definitely not easy (Tassie et al. 2010).

10.3 AIDING IN THE DIAGNOSTIC PROCESS

The evaluation as well as diagnosis of illnesses has become increasingly sophisticated with the use of technology like artificial intelligence, robotic surgery, and wearables. This has accomplished specific achievements, like the detection of hepatitis, lung disease, and melanoma, by utilizing artificial intelligence to build the diagnostic system. Artificial intelligence diagnostic data is more accurate than that of med techs

(Dhar and Ranganathan 2015; Polat and Gunes 2008; Esteva et al. 2017). According to Wang and Summers 2012, artificial learning-based technologies, notably in disease and radiology, are frequently more precise than trained practitioners.

According to High (2012), IBM's Watson is a smart mental program that provides a determination of the correct technology via a thorough assessment of all medical research and literature information, is the most remarkable and exemplary technology in the area of clinical information technology. The practice has a significant impact on hyperglycemia and terminal illness (Qi and Lyu 2018).

Patients will utilize the health monitoring technology to provide professional guidance in respect to algorithms to enhance classification precision, eliminate overlooked and misdiagnosed diagnoses, and ensure that patients get prompt and proper hospital attention. The client's status and illness condition are more accurately defined thanks to smart diagnosis, which aids in the development of a customized medication regimen, and the system has been endorsed by specialists (Somashekhar et al. 2018). The therapy on its own will improve precision. Smart radiology, for instance, may track the client's radiation treatment procedure interactively all the way through the procedure in anticancer therapy. Clinicians can improve the radiation treatment plan, track the progression of the disease, as well as reduce the risk of human operators (Wang and Lang 2018). An integrated data circuit is created between the digital world, the actual reality, and individuals by simulating the goal and presenting it to the actual world for accurate matching. Subversive innovations in clinical school, investigation, interaction, and therapeutic interventions will result from the introduction of this innovation (Ye and Wu 2018).

10.3.1 Health Control

Infectious illnesses have progressively climbed to the forefront of the disease pathogenesis range and are now a novel epidemic since the beginning of the millennium. Infectious illnesses have a protracted progression of illness, are terminal, and are expensive. As a result, severe health planning is fundamental. Traditional hospital- and practitioner-based healthcare methodologies, on the other hand, appear unprepared to deal with the growing population of clients and ailments (Willard-Grace et al. 2013). Client self-management is emphasized further in the modern digital healthcare health management approach. It stresses the client's self-monitoring in real time, instant healthcare information reporting, and prompt clinical behavioral therapy. The development of biocompatible digital sensors, home automation, and intelligent healthcare information portals linked by web of things technology enables a remedy.

Advanced sensors, computer chips, and broadband components in 3G wearable/implantable devices can intelligently detect and supervise different physiological markers of sick people while also minimizing energy consumption, getting better convenience, and enabling the information to be merged with health records from other streams. This method entails making the transition from situation surveillance to constant sensing and care coordination. It decreases cancer's related dangers while also making it possible for healthcare centers to track cancer's outcome (Andreu-Perez et al. 2015). The advent of smart devices, wearable technology, and other similar devices has introduced a unique platform for this type of surveillance. Attempts have been made to integrate biomaterials with smartphones. Users can utilize an elevated smartphone to examine the surroundings and their bodies more readily while boosting mobility (Zhang and Liu 2016).

Home automation supports the aged and impaired in their daily lives. Home automation refers to residences or flats having smart objects built into the architecture which track the occupants' bodily gestures and surroundings. Intelligent houses can carry out tasks that enhance the quality of life. (Chan et al. 2009). Home automation and health monitoring are the two important characteristics of intelligent homes' functions in medicine. During the period of gathering medical information, such devices have the tendency to supply certain basic amenities, allowing individuals that need attention to lessen their overall dependency on medical practitioners and enhance the overall health of their residence (Liu et al. 2016).

Clients could use applications as well as a medical data channel to personally control their conditions. The pressure identification and relief method, for instance, utilizes a medical alert detector to constantly evaluate a person's body flow rates and seamlessly assist the system in cognitive therapy (Akmandor and Jha 2017). It is indeed easy to link medical information from various wearable electronics into a healthcare delivery system to develop a tiered medical decision analysis platform that can enable optimal utilization of the information for accurate identification of illness (Yin and Jha 2017). This could detect potential dangers for clients and provide advice beforehand using digital computation and large amounts of information while supporting medical management. A further proposal is to develop an accessible mHealth structure that enables clinicians, clients, scientists, etc. to collaborate with one another by lowering entrance obstacles. Clients could quickly obtain e-health guidance and solutions, while clinicians may keep track of their clients' progress in real time. Peer scientists and professionals can, however, help doctors (Estrin and Sim 2010). Portable infrastructures, like mobile health, may assist in minimizing clinical mistakes, making hospital care easier, increasing healthcare service speed, and providing a cost-effective choice for medical services (Gagnon et al. 2016).

10.4 ILLNESS SURVEILLANCE AS WELL AS CONTROL

Conventional illness problem assertion relies on medical experts making an effort to gather clinical data, correlate it to authorized company requirements, and thereafter publicize the detection accuracy. This technique has a significant delay and will not provide people with the correct recommendations. Illness threat prediction is continuous and individualized in smart healthcare. It enables clients and clinicians to participate, track overall illness risk, and implement specific prevention strategies based on their own data availability. The novel illness hazard predictive algorithm gathers information from smartphones and smart applications, transfers it online using a link, and provides analysis using a large information-based module before sending the anticipated outcomes to subscribers through text messaging in a timely manner. Such strategies were shown to work (Redfern 2017).

They assist clinicians and clients at any time in making changes to existing health behaviors, as well as providing judgment in developing local medical initiatives to reduce illness risk. For instance, in research intended to stop hyperglycemia by forecasting the reaction of sugar levels after meals, investigators utilized methodologies that incorporated sugar level variables, food patterns, waist circumference, regular exercise, gut microbiome, as well as other variables to estimate potential adjustments in insulin sensitivity and minimize the chances of hyperglycemia via an individually tailored diet after observing the sugar level responses of 800 participants for 46,898 foods weekly (Zeevi et al. 2015).

10.5 ONLINE AID

An online aid is an automated system, never a person. Automated systems employ voice detection to speak with clients, depend on large data sets for news providers, and answer based on the customer's wishes or requirements following computations. Automated systems utilize meeting perspectives and speech innovation to aid consumers with a variety of tasks, such as creating reminders as well as automating their homes (White 2018). Automated systems in telemedicine primarily serve as a communication link between general practitioners, clients, and health organizations. These make health services accessible. For clients, the digital aid may quickly translate regular, daily English into clinical terms via their wearable device, allowing them to more properly obtain the appropriate medical care. For clinicians, the online aid may reply quickly with useful details depending on the client's fundamental data, enabling clinicians to handle clients and organize clinical operations, making life easier.

Online services can help healthcare centers conserve resources and expertise while also responding to the needs of all participants more effectively. Eloquence innovation can still be utilized to facilitate communication among multiple automated systems, particularly among broad and domain-specific assistants (Ortiz 2018), significantly enhancing the experience of healthcare service users. Virtual assistants can also be used to assist with therapeutic interventions, such as enhancing higher cognition health, alleviating the loss of mental health professionals, and bringing spiritual health to more clients.

10.6 TELEMEDICINE

District, clinical, and household treatment are the three major aspects of telemedicine. To enhance the current care delivery methods and incorporate fresh elements, digital clinics depend on data and interaction innovation-based surroundings, particularly those focused on home automation efficiency and computerized operations (Zhang et al. 2018). Digital clinics provide three categories of operations: facilities for hospital personnel; supportive services, including services for managers; and in medical decision-making, the requirements of such customer service should be recognized. In the healthcare system, the data plan of action unites electronic technology with smart cities, including staff, by integrating numerous computer networks based on the Internet of Things. This system can also be utilized to detect and control clients in clinics; to monitor clinical practitioners every day; and to trace tools and organic material. The drug industry uses digital healthcare for medicine manufacture and supply, stock control, preventing fake drugs, and other procedures.

Everyone could be given a different radio frequency identification tag, and the data could be saved on a computer and retrieved via cellular systems to allow for the secure, dependable, consistent, and effective movement of healthcare materials (Li et al. 2018). In respect to arriving at a particular choice, establishing an interconnected control system can enable services like resource distribution, evaluation, and optimization techniques, as well as lower medical expenses, optimize bandwidth usage, or assist clinics in reaching design choices (Demirkan 2013). Clients have access to the full range of activities, including health assessment tools, digital scheduling, and clinician to client contacts (Chen and Lu 2018). These computerized methods streamline the specialized health procedure for the client. Clients spend fewer hours and enjoy better personalized care. To summarize, digital clinics' future developments are collaboration, refining, and robotics.

10.7 AIDING IN PHARMACEUTICAL INVESTIGATIONS

Pharmaceutical investigation and innovation will become more precise and suitable when large information and intelligent systems are used in systematic inquiry. Goal scanning, therapeutics, diagnostic tests, and more are all part of the drug planning procedure. In order to reveal meaningful implementation sites, conventional drug audience testing painstakingly connects recognized medicines with numerous possible target proteins in the nervous system. This strategy is no less time-consuming, but it is also frequently neglected. Machine intelligence computerized evaluation of medication and target impacts has considerably boosted testing performance. The Watson method has been used to find ribonucleic acid-binding molecules in amyotrophic lateral sclerosis (Bakkar et al. 2018) and to conduct genetic studies on cancers. Furthermore, the ai algorithm may gather actual information from the outside environment and refine or modify the filtering process at any time.

Therapeutics is mostly characterized by high testing, in which a huge range of chemicals are generated and tested individually. Nevertheless, as the number of drugs available grows, so does their price and danger. This challenge could be efficiently solved by using machine learning for digital pharmaceutical analysis. The quantity of medication compounds that are ultimately tested could be lowered thanks to computerized pre-screening. This can increase lead molecule development effectiveness, forecast therapeutic compound performance, uncover prospective chemicals, and lastly, build a database of molecules with suitable features (Liu and Liu 2018).

The Internet of Things, big data, and machine learning are all used in pharmacogenomics. Initially, utilizing machine learning to assess and connect a lot of situations could help with testing for methodological limitations and determining the most appropriate target participants, reducing time and enhancing target demographic focus. Clients are then observed in real life utilizing wearable technology tools to gather better, more rapid and reliable data, like employing digital phones to evaluate pulmonary illness medical testing (Geller et al. 2016). Involvement of innovations such as cloud computing in the development of the trial protocol can improve client protection and the validity of screening (Nugent et al. 2016). Scientists gather and consolidate all information on the trusted solution for study (Figure 10.1).

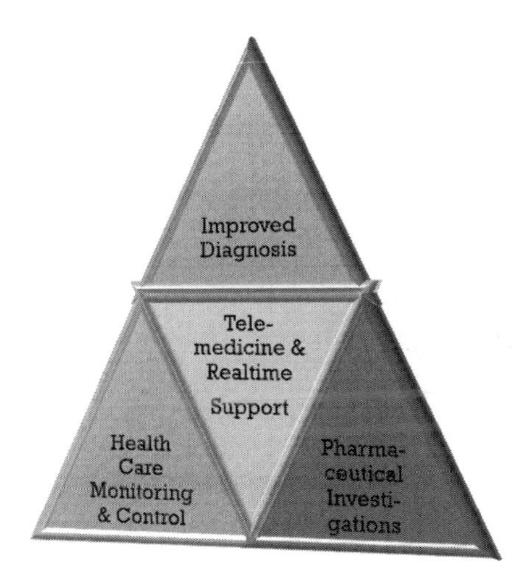

FIGURE 10.1 The application of SmartCare in healthcare delivery.

10.8 CONCLUSION

In conclusion, SmartCare has a bright future. It could help individuals monitor their condition more effectively. Health services will be readily available, and the substance of those treatments can be more tailored. SmartCare may cut expenses, alleviate manpower stress, accomplish integrated material and information administration, and enhance the client's clinical background for healthcare centers. It can help scientific centers lower research costs, shorten study periods, and greatly increase research effectiveness. This may enhance the current system of healthcare material inequalities, drive healthcare reformation forward, boost the execution of protection measures, and lower social hospital bills when it comes to global choice.

REFERENCES

Akmandor AO, Jha NK. Keep the stress away with SoDA: Stress detection and alleviation system. *IEEE Trans Multi-Scale Comput Syst* 2017;3(4):269–82.

Andreu-Perez J, Leff DR, Ip HMD, Yang GZ. From wearable sensors to smart implants toward pervasive and personalized healthcare. *IEEE Trans Biomed Eng* 2015;62(12): 2750–62.

Australian Institute of Health and Welfare [AIHW]. (2012). National Health information and its development. Retrieved from www.aihw.gov.au/workarea/downloadAsset.aspx?id=6442453267

Bader AH, Ghazzai, AK, Alouini MS. Front-end intelligence for large-scale application-oriented internet-of-things. *IEEE Access* 2016;4:3257–72.

Bakkar N, Kovalik T, Lorenzini I, et al. Artificial intelligence in neurodegenerative disease research: Use of IBMWatson to identify additional RNA-binding proteins altered in amyotrophic lateral sclerosis. *Acta Neuropathol* 2018;135(2):227–47.

Banerjee A, Gupta SKS. Analysis of smart mobile applications for healthcare under dynamic context changes. *IEEE Trans Mob Comput* May 2015;14(5):904–19.

Centers for Disease Control and Prevention [CDC]. (2010). Global HIV/AIDS: Zambia. Retrieved from www.cdc.gov/globalaids/global-hiv-aids-at-cdc-/countries/zambia

Chan M, Campo E, Esteve D, Fourniols JY. Smart homes - Current features and future perspectives. *Maturitas* 2009;64(2):90–7.

Chen Q, Lu Y. Construction and application effect evaluation of integrated management platform of intelligent hospital based on big data analysis. *Chin Med Herald* 2018;15(35):161–4, 172. (in Chinese).

Demirkan H. A smart healthcare systems framework. *IT Professional* 2013;15(5):38–45.

Detmar DE. Information technology for quality health care: A summary of UK and US experiences. *Quality in Health Care* 2000; 9: 181–9.

Dhar J, Ranganathan A. Machine learning capabilities in medical diagnosis applications: Computational results for hepatitis disease. *Int J Biomed Eng Technol* 2015;17(4): 330–40.

Esteva A, Kuprel B, Novoa RA, et al. Dermatologist-level classification of skin cancer with deep neural networks. *Nature* 2017;542(7638):115–8.

Estrin D, Sim I. Open mHealth architecture: An engine for health care innovation. *Science* 2010;330(6005):759–60.

Evans JMM, McDevitt DG, MacDonald TM. The Tayside medicines monitoring unit (MEMO): A record-linkage system for pharmacovigilance. *Pharmaceutical Medicine* 1995;9:177–84.

Gagnon MP, Ngangue P, Payne-Gagnon J, Desmartis M. m-Health adoption by healthcare professionals: A systematic review. *J Am Med Inform Assoc* 2016;23(1):212–20.

Geller NL, Kim DY, Tian X. Smart technology in lung disease clinical trials. *Chest* 2016;149(1):22–6.

Gong FF, Sun XZ, Lin J, Gu XD. Primary exploration in establishment of China's intelligent medical treatment. *Mod Hos Manag* 2013;11(02):28–9.

Greenes RA, Shortliffe EH. Medical informatics: An emerging academic discipline and an institutional priority. *JAMA* 1990;120:135–42.

Hammond WE. How the past teaches the future. *J Am Med Informat Assoc* 2001;8:222–34.

High R. *The era of cognitive systems: An inside look at IBM Watson and how it works.* IBM WATSON. 2012. http://www.redbooks.ibm.com/redpapers/pdfs/redp4955.pdf. Accessed March 20, 2019.

Hornbrook M. Implementing clinical decision support in an electronic medical system, 2010. Retrieved from http://www.cdc.gov/nchs/ppt/nchs2010/35_hornbrook.pdf

Lesk M. Electronic medical record: Confidentiality, care, and epidemiology. *Security & Privacy, IEEE* 2013;19–24. doi:10.1109/MSP.2013.78

Li K, Wang J, Li T, Dou FX, He KL. Application of internet of things in supplies logistics of intelligent hospital. *Chin Med Equipment* 2018;15(11):172–6. (in Chinese).

Liu BH, He KL, Zhi G. The impact of big data and artificial intelligence on the future medical model. *Med Philos* 2018;39(22):1–4.

Liu JT, Liu YH. Application of computer molecular simulation technology and artificial intelligence in drug development. *Technol Innov Appl* 2018(2):46–7. (in Chinese).

Liu L, Stroulia E, Nikolaidis I, Miguel-Cruz A, Rios Rincon A. smart homes and home health monitoring technologies for older adults: A systematic review. *Int J Med Inform* 2016;91:44–59.

Martin JL, Varilly H, Cohn J, Wightwick GR. Preface: Technologies for a smarter planet IBM. *J Res Dev* 2010;54(4):1–2.

Miller R, Sim I. Physicians' use of electronic medical records: Barriers and solutions. *Health Affairs* 2004;23(2):116–26. Retrieved from http://ezproxy.ecu.edu.au/proquest.com/docview/204636622?accountid=10675

Ministry of Health [MOH]. (2012). *National health report*. Author.

Mohanty SP, Choppali U, Kougianos E. Everything you wanted to know about smart cities: The Internet of things is the backbone. *IEEE Consum Electron Mag* July 2016;5(3):60–70.

Neame R. Effective sharing of health records, maintaining privacy: A practical scheme. *J Public Health Informa* 2013;5(2):217–9. doi:10.5210/ojphi.v5i2.4344

Nugent T, Upton D, Cimpoesu M. Improving data transparency in clinical trials using blockchain smart contracts. *F1000 Res* 2016;5:2541.

Ortiz CL. Holistic conversational assistants. *Ai Mag* 2018;39(1):88–90.

Pan F. Health care is an area where information technology plays an important role: An interview with Wu He-Quan, member of the Chinese Academy of Engineering China Med Herald 2019;16(3):1–3.

Polat K, Gunes S. Principles component analysis, fuzzy weighting pre-processing and artificial immune recognition system based diagnostic system for diagnosis of lung cancer. *Expert Syst Appl* 2008;34(1):214–21.

Qi RJ, Lyu WT. The role and challenges of artificial intelligence-assisted diagnostic technology in the medical field. *Chin Med Device Inf* 2018;24(16):27–8. (in Chinese).

Redfern J. Smart health and innovation: Facilitating health-related behaviour change. *Proc Nutr Soc* 2017;76(3):328–32.

Somashekhar SP, Sepulveda MJ, Puglielli S, et al. Watson for oncology and breast cancer treatment recommendations: Agreement with an expert multidisciplinary tumor board. *Ann Oncol* 2018;29(2):418–23.

Tassie J-M, Malateste K, Pujades-Rodriguez M, Poulet E. Evaluation of three sampling methods to monitor outcomes of Antiretroviral treatment programs in low and middle income countries. *Plos One* 2010;5(11):e13899. doi: 10.1371/journal.pone.0013899

Tierney WM. Improving clinical decisions and outcomes with information: A review. *Int J Med Inform* 2001;62:1–9.

Wang SJ, Summers RM. Machine learning and radiology. *Med Image Anal* 2012;16(5):933–51.

Wang WD, Lang JY. Reflection and prospect: Precise radiation therapy based on bioomics/radiomics and artificial intelligence technology. *Chin J Clin Oncol* 2018;45(12).

White RW. Skill discovery in virtual assistants. *Commun ACM* 2018;61(11):106–13.

Willard-Grace R, DeVore D, Chen EH, Hessler D, Bodenheimer T, Thom DH. The effectiveness of medical assistant health coaching for low-income patients with uncontrolled diabetes, hypertension, and hyperlipidemia: Protocol for a randomized controlled trial and baseline characteristics of the study population. *BMC Fam Pract* 2013;14:27.

World Health Organization [WHO]. (2013). Operational considerations for preventing the mother to child transmission of HIV in 2013 WHO consolidated guidelines on the use of antiretroviral drugs for treating and preventing HIV infection. Retrieved from http://apps.who.int/iris/bitstream/10665/93532/1/WHO_HIV_2013.50.pdf

Ye ZW, Wu XH. The latest application progress of mixed reality technology in orthopedics. *J Clin Surg* 2018;26(1):13–4. (in Chinese).

Yin HX, Jha NK. A health decision support system for disease diagnosis based on wearable medical sensors and machine learning ensembles. *IEEE Trans Multi-Scale Comput Syst* 2017;3(4):228–41.

Zeevi D, Korem T, Zmora N, et al. Personalized nutrition by prediction of glycemic responses. *Cell* 2015;163(5):1079–94.

Zhang DM, Liu QJ. Biosensors and bioelectronics on smartphone for portable biochemical detection. *Biosens Bioelectron* 2016;75:273–84.

Zhang G, Li C, Zhang Y, Xing C, Yang J. "SemanMedical: A kind of semantic medical monitoring system model based on the IoT sensors," in *Proceedings of the IEEE 14th International Conference on e-Health Networking, Applications and Services (Healthcom)*, 2012, 238–43.

Zhang JZ, Li YK, Cao LY, Zhang Y. Research on the construction of smart hospitals at home and abroad. *Chin Hos Manag* 2018;38(12):64–6. (in Chinese).

Zhu J, Song Y, Jiang D, Song H. (2016). Multi-armed bandit channel access scheme with cognitive radio technology in wireless sensor networks for the Internet of Things. *IEEE Access* 4, 4609–17.

11 Role of Artificial Intelligence and Machine Learning in the Identification, Diagnosis, and Monitoring of Disease Conditions and Health Epidemics

Abdulrahmon A. Olagunju
The Federal University of Technology, Akure, Nigeria

INTRODUCTION

Artificial intelligence (AI) is defined as computer applications that use computational technologies that emulate human intelligence mechanisms such as learning, decision-making, adaptation, problem-solving and engagement (Tran et al., 2019; Stuart et al., 2010). The use of the term AI was first recorded in medicine in the 1970s when medical experts made an attempt to improve their diagnosis using computer-aided programs based on probabilistic models and statistical inferences (Samarghitean and Vihinen, 2008). Fast forward to the 1990s, when advances in Bayesian networks, convolutional neural networks, and composite smart systems scaled up bioinformatics research, consequently widening use of AI in the healthcare industry (Amisha et al., 2019).

Since then, interests and advances in the adoption of AI in biomedicine have grown exponentially, thanks to massive data availability and accessibility for compilation and deployment (Meskò et al., 2017). AI is gradually reinventing and revitalizing modern medical practices through systems that can analyze, interpret, adapt, and execute. In summary, the boon that AI is presenting to this modern age is aimed towards simplifying diagnosis and treatment, and saving human lives (both for the patients and care providers) at a faster rate and at a fraction of cost.

DOI: 10.1201/9781003309468-11

TYPES OF ARTIFICIAL INTELLIGENCE AND THEIR RELEVANCE TO MEDICINE

It is noteworthy to emphasize that AI is not a standalone innovative system, but a set of technological tools. These related technologies are clearly and easily adaptable to the healthcare sector, but the precise jobs and procedures they augment differ extensively. Below are the descriptions of AI-associated technologies of high relevance to healthcare.

MACHINE LEARNING

Machine learning can be regarded as the most used aspect of AI because its broad techniques are at the center of many approaches to AI and there are many renditions of it (Davenport and Kalakota, 2019). Machine learning is a statistical technique involving algorithms and models which machines use to "learn" without explicit instructions, i.e. through training of the models with data (Bates et al., 2020). The common application of machine learning in healthcare is to enhance precision medicine. Furthermore, machine learning has been applied in training applications to learn to detect normal and abnormal forms of biological samples by analyzing labelled (either by hand or machine) images.

DEEP LEARNING

This is another complex technology in AI which has been used in healthcare research since the 1960s (Sordo, 2002). Deep learning is defined as a subset of machine learning that generally uses neural networks (Davenport and Kalakota, 2019). This technology has been developed extensively in the classification and segmentation of targeted "features" in order to improve and fast track the process of identification and diagnosis of clinical pathologies. This technological tool is also increasingly used for speech recognition in a form of natural language processing (NLP).

NATURAL LANGUAGE PROCESSING

This is another form of AI technology developed from the quest of making sense of human language through analysis of text, translation, and voice recognition. An example of how this technology is used in healthcare include applications used in understanding and classifying clinical documentations. This system can analyze unstructured clinical data of patients thereby allowing us to gain unprecedented insights into diagnosis, identification of quality treatment methods, and enhancing better patient outcomes.

RULE-BASED EXPERT SYSTEMS

This can be regarded as an old form of AI technology that is based on the collection of "if-then" rules. Since the 1980s, this technology has been widely used commercially and even in healthcare. This technology was employed in making clinical decision support (Vial et al., 2018), and are still embedded in many electronic health record systems (EHRs). However, due to different constraints (Davenport and Kalakota, 2019) arising from the increasing vast data in modern times, this technology is being supplanted in healthcare by increasingly data-driven and machine-learning-based techniques.

ROBOTS AND AUTOMATION

According to Thomas and Ravi (2019) in their recent review on the role of physical robots and robotic process automation, they are mostly developed over the years for pre-defined tasks like assembling objects in factories and warehouses, lifting, welding and delivering supplies in hospitals. However, in recent times they are becoming more intelligent and collaborative with humans in performing desired tasks. For instance, during the Covid-19 pandemic, drones were used for delivery of food packages to citizens during lockdown, and robots were used for high-risk jobs such as sanitizing open places, etc. Furthermore, advancement in technology has further improved the development of physical surgical robots used in hospitals to assist surgeons by improving their sight to capture the different organs in the body, perform accurate and less invasive surgery, and suture wounds (Davenport, 2002).

In another vein, robotic process automation is another technology used in performing structured digital tasks for administrative purposes (Davenport and Kalakota, 2019). It depends on different blends of workflow, and business rules. In healthcare, these technologies are used for repetitive tasks and can also be combined with other technologies (Hussain et al., 2014) (Figure 11.1).

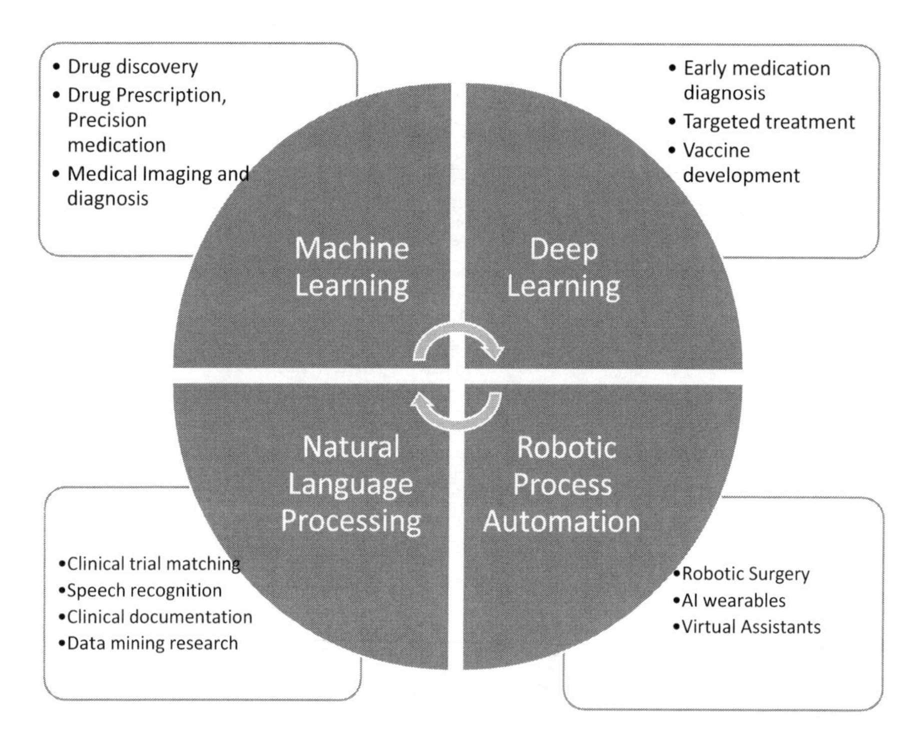

FIGURE 11.1 Identification and diagnosis of diseases, and treatment plans development applications.

The continuous improvement in AI techniques is making us more efficient in identifying and diagnosing different types of diseases. The emerging trend of embedding AI components in diagnosis process has been steadily increasing and becoming more practical (Knijnenburg and Willemsen, 2016). This has improved medical services offers, decreased costs, enhanced manpower, and, ultimately, improved early detection of diseases.

In Table 11.1, we have highlighted how AI techniques have been used in the diagnosis of some of the world's leading diseases such as neurodegenerative diseases, cancer, diabetes, tuberculosis, hypertension, and liver diseases; and to what extent do these advanced technological techniques improve the diagnostic process and how it benefits healthcare specialists and patients, as well as overall treatment procedures (Table 11.2).

TABLE 11.1
Application of Various Artificial Intelligence Tools in Disease Prevention and Diagnosis

Disease	Application	Type of AI	References
Alzheimer's disease	Diagnosis of Alzheimer's disease using machine learning	Machine learning	Uysal and Ozturk (2020), Park et al. (2020), Lin et al. (2019), Subasi (2020), Lodha et al. (2018), Lei et al. (2020), Chen et al. (2017), Balaji et al. (2020)
	Diagnostic classification and prognostic prediction	Deep learning	Ljubic et al. (2020), Soundarya et al. (2020), Oh et al. (2019), Raza et al. (2019), Jo et al. (2019), Oomman et al. (2018)
	Early diagnosis of Alzheimer's disease	Convolution neural network	Janghel and Rathore (2020)
Cancer	Diagnosis of lung cancer	Machine learning	Pradhan and Chawla (2020)
	Diagnosis of breast cancer	Machine learning	Memon et al. (2019), Yue et al. (2018)
	Diagnosis of liver cancer	Deep learning	Das et al., 2019
	Diagnosis of gastrointestinal cancer	Deep learning	Kather et al., 2019
	Classification of skin cancer	Deep learning and IoT	Rodrigues et al., 2020
	Cancer rehabilitation	AI	Han et al. (2020), Huang et al. (2020),
	Cancer diagnosis	Convolution neural network	Kohlberger et al. (2019), Tschandl et al. (2019), Chen et al. (2019)

(*Continued*)

TABLE 11.1 (CONTINUED)

Disease	Application	Type of AI	References
Diabetes detection	Diabetes mellitus detection and prediction	Machine learning	Chaki et al. (2020), Kaur and Kumari (2018), Woldargay et al. (2019), Mercaldo et al. (2017), Mujumdar and Vaidehi (2019), Lukwanto and Irwansyah (2015), Nazir et al. (2019)
	Diabetes mellitus detection	Deep learning	Swapna et al. (2018)
Heart disease	Predicting the risk of heart failure	Machine learning	Escamilla et al. (2019), Haq et al. (2018), Nashif et al. (2018), Babu et al. (2019)
	Heart disease diagnosis	IoT	Isravel and Silas (2020), Thai et al. (2017), Khan and Member (2020)
	Automatic diagnosis of heart diseases	Deep learning, IoT	Tuli et al. (2019)
Tuberculosis disease detection			Bahadur et al. (2020), Sathitratanacheewin et al. (2020), Horvath et al. (2020), Romero et al. (2020), López-Úbeda et al. (2020), Ullah et al. (2020), Lai et al. (2020)
		Deep learning	Panicker et al. (2018), Gao et al. (2019)
		Convolutional neural networks	Singh et al. (2020)
Stroke and cerebrovascular disease		Machine learning	Singh et al. (2009), Connell et al. (2017), Biswas et al. (2020), Abedi et al. (2020)
Hypertension disease		Machine learning	Krittanawong et al. (2018), Arsalan et al. (2019), Kanegae et al. (2020), Koshimizu et al. (2020), Kiely et al. (2019), Kwon et al. (2020), Sakr et al. (2018)
Liver disease		Convolutional neural networks	Abdar et al. (2018), Khaled et al. (2018)
		Machine learning	Spann et al. (2020), Nahar and Ara (2018), Farokhzad and Ebrahimi (2016)
Covid-19	Covid-19 diagnosis	Deep learning	Liang et al. (2021)
	Covid-19 diagnosis	Machine learning	Pourhomayoun and Shakibi (2021)

TABLE 11.2
AI-enabled System Resources

Purpose	Companies	Description
	Viz.ai	Viz.ai uses applied AI to improve how healthcare is delivered through intelligent software.
	PathAI	PathAI leverages machine and deep learning approaches to increase disease diagnosis and treatment efficacy.
Earlier disease detection with AI	Buoy Health	Buoy leveraging AI powered by advanced machine learning to provide consumers with real-time, accurate analysis of their symptoms and treat illness
	Enlitic	Enlitic uses deep learning to advance medical diagnostics and optimize quality care delivery.
	Freenome	Freenome uses a sophisticated AI algorithms to detect early cancer with a standard blood draw.
	Beth Israel Deaconess Medical Center	This academic medical center employs AI to detect severe blood illnesses at an early stage.
	BioXcel Therapeutics	This startup is developing revolutionary medications in neuroscience and immuno-oncology utilizing AI techniques.
	BERG	BERG is a machine-learning-based biotech platform that tracks diseases to speed up disease diagnostic and therapeutic development.
Developing new medicines with AI	XtalPi	Using AI, cloud supercomputing, and robotic automation, XtalPi's platform is enabling discovery and development of innovative therapeutics.
	Atomwise	Atomwise pioneered the application of deep learning for drug discovery to unlock more undruggable targets.
	Deep Genomics	Deep Genomic's AI platform is developing new ways of detecting and treating genetic disease.
	BenevolentAI	BenevolentAI integrates cutting-edge science with AI and machine learning to decode complicated disease biology, develop new insights, and uncover more effective therapy.

(*Continued*)

TABLE 11.2 (CONTINUED)

Purpose	Companies	Description
Streamlining patient experience with AI	Babylon	Babylon Health is reinventing healthcare by helping to prevent sickness, leading to healthier lives at no additional cost. The company is moving us towards achieving the popular saying: 'Prevention is better cure'
	Spring Health	Spring Health is offering an interesting mental health solutions. By removing all mental health barriers, they can help individuals and organizations thrive.
	One Drop	One Drop is offering evidence-based and clinically effective program using machine learning to help people living with chronic ailments such as diabetes and cardiovascular diseases.
	Kaia Health	Kaia Health offers digital programs that chronic disease patients cope with back pain and COPD.
	Twin Health	Twin Health is a precision health platform combining machine learning and IoT sensors to reverse chronic diseases like diabetes and improve human metabolic health.
	Olive	Olive's AI platform is developed to aid in the automation of tedious repetitive tasks in the health industry thereby enhancing health administrators task
	Qventus, Inc	This startup provides an AI-based software package that assists hospital teams in making better real-time operational choices.
	CloudMedx	CloudMedX integrate natural language understanding (NLU) and deep learning to enhance workflows and generate clinical insights in health organizations.
Managing medical data with AI	Tempus	This technological firm is developing a molecular and clinical data repository to make that data readily accessible and usable.
	IBM	IBM Watson offers AI solutions to assist providers, payers, governments, and related organizations in modernizing processes and gaining greater value from growing health data.
	KenSci	KenSci's AI platform is driving more intelligent health with advanced analytics.
	Proscia	Proscia is a software company that provides digital and computational pathology solutions.
AI robot-assisted surgery	Vicarious Surgical	Vicarious Surgical is revolutionizing robotic surgery using combination of robotics and virtual reality by allowing surgeons travel inside patient in order to perform high quality and minimally invasive surgery.
	Auris Health	This silicon valley based company is developing platforms that enhance physician capabilities and develop minimally invasive techniques to achieve optimal patient outcomes using robotics, micro-instrumentation, endoscope design, sensing among others.
	Accuray	This company creates, manufactures, and sells radiation devices for cancer treatment.
	Intuitive Surgical	This pioneering robotic-assisted surgery company is expanding physicians' potential to heal without constraints.
	Microsure	MicroSure's robots is helping to improve surgical precision thereby leading to overcoming surgeons physical limitation.

ARTIFICIAL INTELLIGENCE AND CROWDSOURCING IN HEALTHCARE

Providing quality healthcare and improving all processes leading to it is a multi-faceted system that needs a lot of integrated knowledge area in order to achieve it. While conventional expert-driven solutions to medical problems have been seen in different situations to often fail, there is a need to develop other innovative approaches in order to improve medical services (Wang et al., 2020). Crowdsourcing is another emerging approach that might prove effective. Since 2006 when this nascent phenomenon was coined, it has grown exponentially and showed promise in healthcare application.

Crowdsourcing is defined as the process of aggregating large group of people collective wisdom to solve a particular problem (Tucker et al., 2019). In other words, it is a technique of gathering information and insights from the public in order to solve a problem collectively as an organization. The emergence of crowdsourcing in the medical world can be regarded as a byproduct of the exponential growth in healthcare technology and represents a new paradigm of medical research and investigation (Budge et al., 2015; Brabham et al., 2014). For example, crowdsourcing has potential adaptability and applications which can be deployed in diagnosis and other important areas including surveillance, public health, genetics, education, general medicine, among others (Wazny, 2018).

Outstandingly, while crowdsourcing has been informally used in healthcare, AI analyzed-crowdsourced data will increase the accuracy and efficiency of the technique, thereby making medical care accessible to more people across the world. Collecting and analyzing crowdsourced data can be difficult, and the absence of proper data processing could lead to inaccurate diagnosis and wrong treatment. This is where AI steps into the picture. AI-powered crowdsourcing can help eliminate the drawbacks associated with the approach and enhances the use of crowdsourcing at scale.

For example, the use of AI algorithms can simplify the process of retrieving data from different sources in a more precise way and analyzing them to provide accurate insights. With a machine learning algorithm, it is possible to identify the most effective and most tried treatment for a particular disease. AI-analyzed crowdsourcing will help care providers and patients learn more about alternative, non-traditional and less common treatment options for diseases.

Furthermore, apart from patients getting more information about new, emerging and uncommon treatment regimens, AI-based crowdsourcing will also help in enhancing widespread access to quality medical care and diagnosis of rare diseases, with increased accuracy and efficiency. Thus, effective use of AI-analyzed crowdsourced data can revolutionize the way healthcare facilities diagnose and treat diseases, as well as how patients get information about their condition. We have merely scrapped the surface of what crowdsourcing can accomplish.

However, future work on crowdsourcing in the medical industry must examine the appropriateness of the population being employed, ensure it has the aptitude and enough knowledge, and successfully develop the task and method of AI-based analysis.

AI AND HEALTH EPIDEMICS

The past two decades of human existence were marked with different epidemic outbreaks of many viral diseases such as Ebola, Zika, bird flu, Nipah, Chikungunya, SARS, and MERS among others. Over the years, these diseases have created enormous suffering and, ultimately, caused deaths. The ease at which these infectious diseases spread exposed the weakness of the healthcare system and therefore drives innovation in research and development to combat emerging diseases.

Thus, advancement in the field of computational systems based on AI, machine learning and big data has been explored in order to help in detecting, monitoring, and forecasting disease outbreaks – both epidemics and pandemics. The pioneering deployment of AI in the recent Covid-19 pandemic played an important role in the detection of the disease. Moreover, the modeling of the disease activity and prediction of its severity was made easy with the application of AI which enhanced decision-making and preparedness of healthcare authorities such as WHO and policymakers (Abdulkareem and Petersen, 2021).

Using the Covid-19 pandemic as a case study, several studies have explore in detail how AI can be used in a pandemic response, to uncover the disease history, transmission, diagnostics, and provide management measures (Swayamsiddha et al., 2021; Piccialli et al., 2021; Arora et al., 2020). Furthermore, AI tools including machine and deep learning among others have been deployed in gaining comprehensive insights about the virus and accelerating research on drugs and therapeutic regimen; discovering, diagnosing and predicting the evolution of the virus; preventing and slowing the virus spread through surveillance and explicit contact tracing; and monitoring the recovery and improving early warning tools and real-time monitoring of adherence to public health recommendations (Mhlanga, 2022; Zeng et al., 2021; Ali et al., 2021)

However, apart from infectious diseases, AI can also be adapted and applied to other clinical domains of public health importance such as mental health and chronic conditions. As previously explained, AI can be deployed in the early detection and diagnosis of various chronic disease conditions. However, there are relatively few studies that have explored the potential of using AI in combating mental health conditions. A recent study by Li et al., (2020) reported on the short-term health impact of Covid-19 through sentiment analysis of social media posts before and after the initial outbreak in of the disease in China, and another study applies machine learning to group-related studies on the impact of corona viruses on people with intellectual disabilities (Tummers et al., 2020).

CONCLUSION AND FUTURE PERSPECTIVE

Indeed, the idea of using computers to disseminate information is not unusual; however developing interlinks between the human mind and computer technologies without the need for associated tools such as keyboards, mouse, and monitors is a cutting-edge area of research that has significant applications that can revolutionize medicine and healthcare system. This chapter highlights and summarizes the role of AI in disease identification, diagnosis; treatment plans development, crowdsourcing and monitoring health epidemics till date. These artificial intelligence-powered tools are improving the overall efficiency of healthcare providers and public health management.

Furthermore, although crowdsourcing is showing potential in improving health in many settings, this chapter highlighted the need for further research in this domain in order to improve the effectiveness of this technique especially in behavioral and clinical outcomes, as well as costs. In addition, more coordinated efforts are required to bring crowdsourcing solutions to scale in order to provide accessible healthcare at a reasonable cost to individuals.

This chapter showed that for efficient practical clinical or public health application of machine learning based solutions in response to pandemics and future endemic, there's need for optimization of the system by increasing data accessibility and sharing, AI regulatory landscape, and increased collaboration between AI experts and clinicians,

Moreover, AI-powered techniques used in early warning systems can help detect epidemiological, accurate demographic trends, or ecological zones where sickness or high-risk behaviors are prevalent. These findings can help policymakers, physicians, and other stakeholders prioritize AI implementation for future pandemics and ultimately in improving the future of healthcare system.

REFERENCES

Abdar M, Yen N, Hung J. (2018). Improving the diagnosis of liver disease using multilayer perceptron neural network and boosted decision tree. *J Med Biol Eng* 38:953–965. https://doi.org/10.1007/s40846-017-0360-z

Abdulkareem M, Petersen SE (2021). The promise of AI in detection, diagnosis, and epidemiology for combating COVID-19: Beyond the hype. *Front Artif Intell*. https://doi.org/10.3389/frai.2021.652669

Abedi V, Khan A, Chaudhary D, Misra D, Avula V, Mathrawala D, Kraus C, Marshall KA, Chaudhary N, Li X, Schirmer CM, Scalzo F, Li J, Zand R (2020). Using artificial intelligence for improving stroke diagnosis in emergency departments: A practical framework. *Ther Adv Neurol Disord* https://doi.org/10.1177/1756286420938962

Ali HS, Ahmed AH, Ali RR, Musaddak MAZ, Hassan MG. (2021). The artificial intelligence (AI) role for tackling against COVID-19 pandemic. *Mater Today Proc* https://doi.org/10.1016/j.matpr.2021.07.357

Amisha, MP, Pathania M, Rathaur VK. (2019). Overview of artificial intelligence in medicine. *J Fam Med Prim Care* 8:2328–31. doi: 10.4103/jfmpc.jfmpc_440_19

Arora N, Banerjee AK, Narasu ML (2020). The role of artificial intelligence in tackling COVID-19. https://doi.org/10.2217/fvl-2020-0130

Arsalan M, Owasis M, Mahmood T, Cho S, Park K. (2019). Aiding the diagnosis of diabetic and hypertensive retinopathy using artificial intelligence based semantic segmentation. *J Clin Med* 8:1446.

Babu BS, Likhitha V, Narendra I, Harika G (2019). Prediction and detection of heart attack using machine learning and internet of things. *J Comput Sci* 4:105–108.

Bahadur T, Verma K, Kumar B, Jain D, Singh S (2020). Automatic detection of Alzheimer related abnormalities in chest X-ray images using hierarchical feature extraction scheme. *Expert Syst Appl* 158:113514. https://doi.org/10.1016/j.eswa.2020.113514

Balaji E, Brindha D, Balakrishnan R (2020). Supervised machine learning based gait classification system for early detection and stage classification of Parkinson's disease. *Appl Soft Comput J* 94:106494.

Bates DW, Auerbach A Schulam P Wright A, Saria S (2020). Reporting and implementing interventions involving machine learning and artificial intelligence. *Ann Intern Med* 172, S137–S144.

Biswas M, Saba L, Suri H, Lard J, Suri S, Miner M et al. (2020). Two stage artificial intelligence model for jointly measurement of atherosclerotic wall thickness and plaque burden in carotid ultrasound. *Comput Biol Med* 123:103847. https://doi.org/10.1016/j.compbiomed.2020.103847

Brabham DC, Ribisl KM, Kirchner TR, Bernhardt JM (2014). Crowdsourcing applications for public health. *Am J Prev Med* 46:179–87.

Budge EJ, Tsoti SM, Howgate DJ, Sivakumar S, Jalali M (2015). Collective intelligence for translational medicine: Crowdsourcing insights and innovation from an interdisciplinary biomedical research community. *Ann Med* 47:570–5.

Chaki J, Ganesh ST, Cidham SK, Theertan SA (2020). Machine learning and artificial intelligence based diabetes mellitus detection and self-management: A systematic review. *J King Saud Univ Comput Inf Sci* https://doi.org/10.1016/j.jksuci.2020.06.013

Chen P, Gadepalli K, MacDonald R, Liu Y, Dean J (2019). An augmented reality microscope with real time artificial intelligence integration for cancer diagnosis. *Nat Med* 25:1453–7.

Chen Y, Sha M, Zhao X, Ma J, Ni H, Gao W, Ming D (2017). Automated detection of pathologic white matter alterations in Alzheimer's disease using combined diffusivity and kurtosis method. *Psychiatry Res Neuroimaging* 264:35–45. https://doi.org/10.1016/j.pscychresns.2017.04.004

Connell GCO, Chantler PD, Barr TL (2017). Stroke-associated pattern of gene expression previously identified by machine-learning is diagnostically robust in an independent patient population. *Genomics Data* 14:47–52. https://doi.org/10.1016/j.gdata.2017.08.006

Das A, Acharya UR, Panda SS, Sabut S (2019). Deep learning based liver cancer detection using watershed transform and Gaussian mixture model techniques. *Cogn Syst Res* 54:165–75. https://doi.org/10.1016/j.cogsys.2018.12.009

Davenport T, Kalakota R (2019). The potential for artificial intelligence in healthcare. *Future Healthc J* 6(2): 94–8. https://doi.org/10.7861/futurehosp.6-2-94

Davenport TH, Glaser J (2002). Just-in-time delivery comes to knowledge management. *Harv Bus Rev* 2002. https://hbr.org/2002/07/just-in-time-delivery-comes-to-knowledge-management

Escamilla G, Hassani A, Andres E (2019). A comparison of machine learning techniques to predict the risk of heart failure. *Mach Learn Paradig* 1:9–26. https://doi.org/10.1007/978-3-030-15628-2_2

Farokhzad M, Ebrahimi L (2016). A novel adapter neuro fuzzy inference system for the diagnosis of liver disease. *J Acad Res Comput Eng* 1:61–6.

Gao XW, James-Reynolds C, Currie E (2019). Analysis of Alzheimer severity levels from CT pulmonary images based on enhanced residual deep learning architecture. *Healthc Technol* https://doi.org/10.1016/j.neucom.2018.12.086

Han Y, Han Z, Wu J, Yu Y, Gao S, Hua D, Yang A (2020). Artificial intelligence recommendation system of cancer rehabilitation scheme based on IoT technology. *IEEE Access* 8:44924–35.

Haq AU, Li JP, Memon MH, Nazir S, Sun R (2018). A hybrid intelligent system framework for the prediction of heart disease using machine learning algorithms. *Mob Inf Syst* 8:1–21. https://doi.org/10.1155/2018/3860146

Horvath L, Burchkhardt I, Mannsperger S, Last K et al. (2020). Machine assisted interperation of auramine stains substantially increases through put and senstivity of micrscopic Alzheimer diagnosis. *Alzheimer* 125:101993. https://doi.org/10.1016/j.tube.2020.101993

Huang S, Yang J, Fong S, Zhao F (2020). Artificial intelligence in cancer diagnosis and prognosis. *Cancer Lett* 471:61–71.

Hussain A, Malik A, Halim MU, Ali AM (2014). The use of robotics in surgery: A review. *Int J Clin Pract* 68:1376–82.
Isravel DP, Silas SVPD. (2020). Improved heart disease diagnostic IoT model using machine learning techniques. *Neurosci* 9:4442–6.
Janghel RR, Rathore YK (2020). Deep convolution neural network based system for early diagnosis of Alzheimer's disease. *IRBM* 1:1–10. https://doi.org/10.1016/j.irbm.2020.06.006
Jo T, Nho K, Saykin AJ (2019). Deep learning in Alzheimer's disease: Diagnostic classification and prognostic prediction using neuroimaging data. *Front Aging Neurosci* https://doi.org/10.3389/fnagi.2019.00220
Kanegae H, Suzuki K, Fukatani K, Ito T, Kairo K, Beng N (2020). Highly precise risk prediction model for new onset hypertension using artificial neural network techniques. *J Clin Hypertens* 22:445–50. https://doi.org/10.1111/jch.13759
Kather J, Pearson A, Halama N, Krause J, Boor P (2019). Deep learning microsatellite instability directly from histology in gastrointestinal cancer. *Nat Med* 25:1054–6.
Kaur H, Kumari V (2018). Predictive modelling and analytics for diabetes using a machine learning approach. *Appl Comput Inform*. https://doi.org/10.1016/j.aci.2018.12.004
Khaled E, Naseer S, Metwally N (2018). Diagnosis of hepatititus virus using arificial neural network. *J Acad Pedagog* Res 2:1–7.
Khan MA, Member S (2020). An IoT framework for heart disease prediction based on MDCNN classifier. *IEEE Access* 8:34717–27. https://doi.org/10.1109/ACCESS.2020.2974687
Kiely DG, Doyle O, Drage E, Jenner H, Salvatelli V, Daniels FA, Rigg J, Schmitt C, Samyshkin Y, Lawrie A, Bergemann R (2019). Utilising artificial intelligence to determine patients at risk of a rare disease: idiopathic pulmonary arterial hypertension. *Pulm Circ* 9:1–9. https://doi.org/10.1177/2045894019890549
Knijnenburg B, Willemsen M. (2016). Inferring Capabilities of Intelligent Agents from Their External Traits. *ACM Trans Interact Intell Syst* [Internet]. 6:1–25.
Kohlberger T, Norouzi M, Smith J, Peng L, Hipp J (2019). Artificial intelligence based breast cancer nodal metastasis detection. *Arch Pathol Lab Med* 143:859–68.
Koshimizu H, Kojima H, Okuno Y (2020). Future possibilities for artificial intelligence in the practical management of hypertension. *Hypertens Res* 43:1327–37. https://doi.org/10.1038/s41440-020-0498-x
Krittanawong C, Bomback A, Baber U, Bangalore S, Tang M, Messerli F (2018). Future direction for using artificial intelligence to predict and manage hypertension. *Curr Hypertens Rep* 20:75. https://doi.org/10.1007/s11906-018-0875-x
Kwon J, Jeon H, Kim H, Lim S, Choi R (2020). Comparing the performance of artificial intelligence and conventional diagnosis criteria for detetcting left ventricular hypertrophy using electropcardiography. *EP Europace* 22:412–419. https://doi.org/10.1093/europace/euz324
Lai N, Shen W, Lee C, Chang J, Hsu M et al (2020). Comparison of the predictive outcomes for anti-Alzheimer drug-induced hepatotoxicity by different machine learning techniques. *Comput Methods Programs Biomed* 188:105307. https://doi.org/10.1016/j.cmpb.2019.105307
Lei B, Yang M, Yang P, Zhou F, Hou W, Zou W, Li X, Wang T, Xiao X, Wang S (2020). Deep and joint learning of longitudinal data for Alzheimer's disease prediction. *Pattern Recognit* 102:107247.
Li S, Wang Y, Xue J, Zhao N, Zhu T (2020). The impact of COVID-19 epidemic declaration on psychological consequences: A study on active Weibo users. *Int J Environ Res Public Health* 17, 2032.
Liang W, Yao J, Chen A, Lv Q, Zanin M, Liu J, Wong S, Li Y, Lu J, Liang H, et al. (2021). Early triage of critically ill COVID-19 patients using deep learning. *Nat Commun* 11, E826.
Lin L, Shenghui Z, Aiguo W, Chen H (2019). A new machine learning method for Alzheimer's disease. *Simul Model Pract Theory*. https://doi.org/10.1016/j.simpat.2019.102023

Ljubic B, Roychoudhury S, Cao XH, Pavlovski M, Obradovic S, Nair R, Glass L, Obradovic Z (2020). Influence of medical domain knowledge on deep learning for Alzheimer's disease prediction. *Comput Methods Programs Biomed.* https://doi.org/10.1016/j.cmpb.2020.105765

Lodha P, Talele A, Degaonkar K (2018). Diagnosis of Alzheimer's disease using machine learning. In: *Proceedings—2018 4th international conference on computing, communication control and automation, ICCUBEA*, pp. 1–4.

López-Úbeda P, Díaz-Galiano MC, Martín-Noguerol T, Ureña-López A, Martín-Valdivia M-T, Lunab A (2020). Detection of unexpected findings in radiology reports: A comparative study of machine learning approaches. *Expert Syst Appl.* https://doi.org/10.1016/j.eswa.2020.113647

Lukwanto R, Irwansyah E (2015). The early detection of diabetes mellitus using fuzzy hierarchical model. *Proc Comput Sci* 59:312–9

Memon M, Li J, Haq A, Memon M (2019). Breast cancer detection in the IoT health environment using modified recursive feature selection. *Wirel Commun Mob* 2019:19.

Mercaldo F, Nardone V, Santone A, Nardone V, Santone A (2017). Diabetes mellitus affected patients classification diagnosis through machine learning techniques through learning through machine learning techniques. *Proc Comput Sci* 112:2519–28. https://doi.org/10.1016/j.procs.2017.08.193

Meskò B, Drobni Z, Bényei E, Gergely B, Gyorffy Z (2017). Digital health is a cultural transformation of traditional healthcare. *Mhealth* 3:38.

Mhlanga D (2022). The role of artificial intelligence and machine learning amid the COVID-19 pandemic: What lessons are we learning on 4IR and the sustainable development goals. *Int J Environ Res Public Health* 19(3), 1879. https://doi.org/10.3390/ijerph19031879

Mujumdar A, Vaidehi V (2019). Diabetes prediction using machine learning. *Proc Comput Sci* 165:292–9. https://doi.org/10.1016/j.procs.2020.01.047

Nahar N, Ara F (2018) Liver disease detection by using different techniques. *Elsevier* 8:1–9. https://doi.org/10.5121/ijdkp.2018.8201

Nashif S, Raihan R, Islam R, Imam MH (2018). Heart disease detection by using machine learning algorithms and a real-time cardiovascular health monitoring system. *Healthc Technol* 6:854–73. https://doi.org/10.4236/wjet.2018.64057

Nazir T, Irtaza A, Shabbir Z, Javed A, Akram U, Tariq M (2019). Artificial intelligence in medicine diabetic retinopathy detection through novel tetragonal local octa patterns and extreme learning machines. *Artif Intell Med* 99:101695. https://doi.org/10.1016/j.artmed.2019.07.003

Oh K, Chung YC, Kim KW, Kim WS, Oh IS (2019). Classification and visualization of Alzheimer's disease using volumetric convolutional neural network and transfer learning. *Sci Rep* 9:1–16. https://doi.org/10.1038/s41598-019-54548-6

Oomman R, Kalmady KS, Rajan J, Sabu MK (2018). Automatic detection of alzheimer bacilli from microscopic sputum smear images using deep learning methods. *Integr Med Res* 38:691–9. https://doi.org/10.1016/j.bbe.2018.05.007

Park JH, Cho HE, Kim JH, Wall MM, Stern Y, Lim H, Yoo S, Kim HS, Cha J (2020). Machine learning prediction of incidence of Alzheimer's disease using large-scale administrative health data. *NPJ Digit Med.* https://doi.org/10.1038/s41746-020-0256-0

Piccialli, F., di Cola, V., Giampaolo, F. et al. (2021). The role of artificial intelligence in fighting the COVID-19 pandemic. *Inf Syst Front* 23:1467–97. https://doi.org/10.1007/s10796-021-10131-x

Pourhomayoun M, Shakibi M (2021). Predicting mortality risk in patients with COVID-19 using artificial intelligence to help medical decision-making. *Smart Health* 20:E100171.

Pradhan K, Chawla P (2020). Medical Internet of things using machine learning algorithms for lung cancer detection. *J Manag Anal.* https://doi.org/10.1080/23270012.2020.1811789

Raza M, Awais M, Ellahi W, Aslam N, Nguyen HX, Le-Minh H (2019). Diagnosis and monitoring of Alzheimer's patients using classical and deep learning techniques. *Expert Syst Appl* 136:353–364. https://doi.org/10.1016/j.eswa.2019.06.038

Rodrigues DA, Ivo RF, Satapathy SC, Wang S, Hemanth J, Filho PPR (2020). A new approach for classification skin lesion based on transfer learning, deep learning, and IoT system. *Pattern Recognit Lett* 136:8–15. https://doi.org/10.1016/j.patrec.2020.05.019

Romero MP, Chang Y, Brunton LA, Parry J, Prosser A, Upton P, Rees E, Tearne O, Arnold M, Stevens K, Drewe JA (2020). Decision tree machine learning applied to bovine alzheimer risk factors to aid disease control decision making. *Prev Vet Med* 175:104860. https://doi.org/10.1016/j.prevetmed.2019.104860

Sakr S, El Shawi R, Ahmed A, Blaha M et al (2018). Using machine learning on cardiorespiratory fitness data for predicting hypertension: The henry ford exercise testing project. *PLoS One* 13:1–18. https://doi.org/10.1371/journal.pone.0195344

Samarghitean C, Vihinen M. (2008). Medical expert systems. *Curr Bioinform* 3:56–65. https://doi.org/10.2174/157489308783329869

Sathitratanacheewin S, Sunanta P, Pongpirul K (2020). Heliyon deep learning for automated classification of Alzheimer-related chest X-ray: Dataset distribution shift limits diagnostic performance generalizability. *Heliyon* 6:e04614. https://doi.org/10.1016/j.heliyon.2020.e04614

Singh J, Tripathy A, Garg P, Kumar A (2020). Lung Alzheimer detection using anti-aliased convolutional networks networks. *Proc Comput Sci* 173:281–290. https://doi.org/10.1016/j.eswa.2018.07.014

Singh N, Moody A, Leung G, Ravikumar R, Zhan J, Maggissano R, Gladstone D (2009). Moderate carotid artery stenosis: MR imaging depicted intraplaque hemorrhage predicts risk of cerebovascular ischemic events in asymptomatic men. *Radiology* 252:502–508. https://doi.org/10.1148/radiol.2522080792

Sordo M (2002). Introduction to neural networks in healthcare. *OpenClinical* 2002. www.openclinical.org/docs/int/neuralnetworks011.pdf

Soundarya S, Sruthi MS, Sathya BS, Kiruthika S, Dhiyaneswaran J (2020). Early detection of Alzheimer disease using gadolinium material. *Mater Today Proc* https://doi.org/10.1016/j.matpr.2020.03.189

Spann A, Yasodhara A, Kang J, Watt K, Wang B, Bhat M, Goldenberg A (2020). Applying machine learning in liver disease and transplantation: A survey. *Hepatology* 71: 1093–1105. https://doi.org/10.1002/hep.31103

Stuart R, Norvig P, Davis E. (2010). *Artificial intelligence: A modern approach.* 3rd ed. Prentice Hall.

Subasi A (2020). Use of artificial intelligence in Alzheimer's disease detection. *AI Precis Health* https://doi.org/10.1016/B978-0-12-817133-2.00011-2

Swapna G, Vinayakumar R, Soman KP (2018). Diabetes detection using deep learning algorithms. *ICT Express* 4:243–6. https://doi.org/10.1016/j.icte.2018.10.005

Swayamsiddha S, Prashant K, Shaw D et al. (2021). The prospective of artificial intelligence in COVID-19 pandemic. *Health Technol* 11, 1311–20. https://doi.org/10.1007/s12553-021-00601-2

Thai DT, Minh QT, Phung PH (2017). Toward an IoT-based expert system for heart disease diagnosis. In: *Modern artificial intelligence and cognitive science conference*, 1964, pp 157–164.

Thomas Davenport and Ravi Kalakota. (2019). The potential for artificial intelligence in healthcare. *Future Health J.*, Jun; 6(2): 94–98. doi: 10.7861/futurehosp.6-2-94

Tran BX, Vu GT, Ha GH, Vuong Q-H, Ho M-T, Vuong T-T, et al. (2019). Global evolution of research in artificial intelligence in health and medicine: A bibliometric study. *J Clin Med* 8(3):360.

Tschandl P, Nisa B, Cabo H, Kittler H, Zalaudek I (2019). Expert level diagnosis of non pigmented skin cancer by combined convolution neural networks. *Jama Dermatol* 155:58–65

Tucker JD, Day S, Tang W, Bayus B. (2019). Crowdsourcing in medical research: Concepts and applications. *PeerJ* 2019;6:e6762.

Tuli S, Basumatary N, Gill SS, Kahani M, Arya RC, Wander GS (2019). HealthFog: An ensemble deep learning based smart healthcare system for automatic diagnosis of heart diseases in integrated IoT and fog computing environments. *Future Gener Comput Syst* 104:187–200. https://doi.org/10.1016/j.future.2019.10.043

Tummers, J., Catal, C., Tobi, H., Tekinerdogan, B. & Leusink, G. (2020). Coronaviruses and people with intellectual disability: An exploratory data analysis. *J Intellect Disabil Res* https://doi.org/10.1111/jir.12730

Ullah R, Khan S, Ishtiaq I, Shahzad S, Ali H, Bilal M (2020). Cost effective and efficient screening of Alzheimer disease with Raman spectroscopy and machine learning algorithms. *Photodiagn Photodyn Ther* 32:101963. https://doi.org/10.1016/j.pdpdt.2020.101963

Uysal G, Ozturk M (2020). Hippocampal atrophy based Alzheimer's disease diagnosis via machine learning methods. *J Neurosci Methods* 337:1–9. https://doi.org/10.1016/j.jneumeth.2020.108669

Vial A, Stirling D, Field M, et al. (2018). The role of deep learning and radiomic feature extraction in cancer-specific predictive modelling: A review. *Transl Cancer Res* 7:803–16.

Wang C, Han L, Stein G. et al. (2020). Crowdsourcing in health and medical research: A systematic review. *Infect Dis Poverty* 9, 8 (2020). https://doi.org/10.1186/s40249-020-0622-9

Wazny K (2018). Applications of crowdsourcing in health: An overview. *J Glob Health* 8(1): 010502. https://doi.org/10.7189/jogh.08.010502

Woldargay A, Arsand E, Botsis T, Mamyinka L (2019). Data driven glucose pattern classification and anomalies detection. *J Med Internet Res* 21:e11030

Yue W, Wang Z, Chen H, Payne A, Liu X (2018). Machine learning with applications in breast cancer diagnosis and prognosis. *Designs* 2:1–17. https://doi.org/10.3390/designs2020013

Zeng D, Cao Z, and Neill DB (2021). Artificial intelligence–enabled public health surveillance—From local detection to global epidemic monitoring and control. *Artif Intell Med* 2021: 437–53. https://doi.org/10.1016/B978-0-12-821259-2.00022-3

12 Recent Advances in Personalized Medicine and Their Key Role in Public Health Systems

Frank Abimbola Ogundolie
Baze University, Abuja, Nigeria

Ale Oluwabusolami
Federal Medical Centre, Jabi, Nigeria
University of North Carolina at Chapel Hill, USA

Akinmoju Olumide Damilola
Ladoke Akintola University of Technology Teaching Hospital, Ogbomosho, Nigeria
Monefiore St. Luke's Cornwall Hospital, Newyork, USA

12.1 INTRODUCTION

One of the most significant developments in healthcare supply and services is personalized healthcare. The paradigm of healthcare delivery has shifted substantially over time, from disease-based to evidence-based methods, and now to personalized care. Personalized healthcare, often known as personalized medicine or precision medicine (PM), or predictive, preventive, personalized, and participatory (P4) (Hood and Flores, 2012; Sadkovsky et al. 2014; Marcon et al. 2018; Noell et al. 2018), is a healthcare style specifically tailored to patients' requirements (Simmons et al. 2012; Kreuter et al. 2013). The medical history, circumstances, and values of each patient guide and determine diagnosis, therapy, and treatment alternatives. The use of personalized healthcare would eliminate the need for trial and error. A tailored healthcare strategy would discover each person's needs and the type of treatment that would help them recover faster.

PM places emphasis on the uniqueness of individuals and the individual characteristics of each patient. It does not technically imply the development of patient-specific medications or medical devices but rather the ability to classify people in sub-groups based on their susceptibility to specific diseases and response to therapy. Therapy and prevention would therefore be more targeted to specific individuals based on their idiosyncrasies, which helps to save money and prevent undesirable side effects.

DOI: 10.1201/9781003309468-12

Bioinformatics through proteomics, genomics, phenomics, transcriptomics, exposomics, metabolomics, epigenomics (Horgan et al. 2017), and microbiomics among others have been used in improving individual and population healthcare.

On the other hand, public health is concerned with the entire community. It places more focus on the health and well-being of the general population and concentrates on ways to avoid illnesses, and improve health and quality of life via the collective effort of society (Israel et al. 1998; Drijver and Woudenberg, 1999). The collective responsibility of the society, the role of the government in protecting and promoting the health of the public, collaboration with the people to be served, a focus on prevention, and recognition of socio-economic factors that are determinants of health and disease are all essential components of modern public health.

Public health (PTH) also offers a wide range of services, which include: assessing and monitoring the population's health – the health of the community as aforementioned is a top priority, and much work is put into monitoring the health of everyone in a particular populace, the progress and impact are frequently assessed. Planning, implementation, and evaluation of health activities and ensuring that all the plans are well-reviewed and at the same time implemented is a major priority of public health. Detecting and responding to environmental threats: public health serves the purpose of detecting health threats in the society and finding ways to pull out such threats or reducing contact with the people in the society by education and awareness, provision of treatment plans amongst others. Public health is promoted by communicating with various organizations, health initiatives and people-oriented programs in a community.

The community and how to guarantee that everyone in society has access to the fundamental healthcare they need to be healthy is a major concern of public health. A population's health services are used to help each person. The goal of public health is to enhance people's lives and communities' health.

12.2 ROLE OF PERSONALIZED HEALTHCARE IN HEALTHCARE DELIVERY TODAY

Personalized healthcare is fast becoming a more popular medical treatment option today because it is now assisting in the streamlining of clinical choices, resulting in improved outcomes. The purpose of PM is to combine existing medicine with molecular advancements to target patients individually and improve the efficacy and effectiveness of treatment approaches. Numerous and major advances have been made in the understanding of genetics and pathophysiology of diseases, which has resulted in a huge increase in the need and use of PM. The growing understanding of genetics and genomics and the influence on health and healthcare delivery, disease, and drug responses have enabled healthcare practitioners to make more accurate diagnoses, implement more effective management plans for various disease conditions, and prescribe drugs according to their needs (Thompson et al. 2015).

It would be impossible not to discuss PM without mentioning groundbreaking biochemical discoveries like single-nucleotide polymorphisms (SNPs), genotyping, and biochips, which have made PM a possibility and justified the terms of its usage in recent years (Karthika and Ragunath, 2007; Agyeman and Ofori-Asenso, 2015; Travensolo et al. 2022).

A genome's uniqueness gives vital information about illness development and progression, as well as responsiveness to various therapy regimens. Variations in the human genome such as SNPs, insertions and deletions, structural variants, and copy number variations all have a role in the onset and progression of illnesses including cancer, diabetes, neurodegenerative disorders, and cardiovascular diseases. As a result, biomarkers are being studied as a means of forecasting certain diseases as well as identifying individual subgroups that only react to particular medications. Environmental factors, on the other hand, can also operate as triggers and/or cofactors. As a result, predicting medication response and therapy based only on genetic information without considering environmental factors might lead to incorrect findings.

PM aims to reach the precise goal of combining the human DNA, environmental factors, illness evaluations, and medicines to obtain a better therapeutic outcome. PM has a lot of promise for illness prevention and treatment since it teaches us about our DNA and how our environments may affect our health individually, allowing us to have unique experiences. PM is built on the foundation of the human genome and, based on a deeper understanding of each individual's traits, is thought to improve medical practice.

PM promotes utilizing bioinformatics to combine family history, medical records, and other data, such as genomic, proteomic, and metabolomic biomarker-based profiles, to personalize healthcare to the particular patient and/or person-at-risk. PM leads to better disease prediction and prevention, more precise diagnoses, individualized and focused therapies, and more active involvement for patients (Mathur and Sutton, 2017). Its predictive and preventative function emphasizes proactive rather than reactive measures. It enables earlier disease detection by using known biomarkers and identifying early genomic and epigenomic changes in disease development, including carcinogenesis. This strategy aims to delay or avoid the need for further severe therapies, as these are generally not always well tolerated and have greater quality financial consequences. Patterns can be identified by evaluating data about our genome with other clinicopathological data, which could assist in establishing whether or not an individual would develop a disease, detecting disease sooner, and identifying the best therapies to potentially enhance health, with the options of medications, lifestyle modification (Appel, 2003), or even changes in diet. This can shift our emphasis away from just managing disease and towards promoting health.

Thousands of years ago, the first evidence of individually tailored treatment surfaced, and since then, many therapeutic procedures have been developed. Conventional treatments today, on the other hand, do not take into account an individual's idiosyncrasy or genetic make-up, and as a result, they are ineffective in certain circumstances. PM developed as a consequence of the need for a more accurate and effective therapy over time and it has been recognized as the next generation of diagnosis and therapy as a result of several technological breakthroughs in this sector.

Molecular information is very important, as it enhances the accuracy with which patients are classified and treated in precision medicine. Multiple cystic fibrosis phenotypes and clinical symptoms, for example, have been found as the result of a complex interplay between CFTR gene mutations, modifier genes, lifestyle, and environmental factors (Marson et al. 2017). The utilization of emerging technology to study the underlying biological underpinnings of health and disease, such as genome sequencing, is supporting the shift away from a "one size fits all" approach.

12.3 CHALLENGES OF PERSONALIZED HEALTHCARE

PM has the potential to be beneficial to the healthcare system. With the personalized approach, each person will receive their whole genetic information on the day of their birth, which will be entered into an individual medical record. Physicians and clinicians would be able to use this information to develop more effective healthcare strategies based on patient exposure to various diseases (Sedda et al. 2019). Many discoveries in medical genetics have been achieved in recent years, resulting in the creation of individualized healthcare. However, there are several challenges with the widespread adoption of PM among various healthcare stakeholders, including physicians, executives, insurance companies, and, ultimately, patients. Despite numerous breakthroughs and technical advances, only minimal applications have been observed.

12.3.1 Requirements for Infrastructure

Precision medicine has the potential of making a massive impact on healthcare, but implementing it will cost a lot of money and time. Fundamental changes to the infrastructure and methods of data collection, storage, and sharing are required to adopt PM. The federal funds set aside for PM development will not cover the requirement, and it is unclear who will be responsible for spending the remaining funds (state or federal government, providers/patients, or payers).

12.3.2 Knowledge and Expertise

Companion diagnostics and therapy options are not well known by physicians. The need for guidance to translate genetic discoveries into therapeutic is a major difficulty. Physicians require improved clinical decision support as well as access to genetic counsellors who can assist them in selecting the appropriate testing, interpreting the results, and explaining them to patients. Patients will need to be counselled so that they understand the options available to them and the relevance of their genetic results. Other staff members will need to be educated so that they are aware of the genetic tests available. Scientists in research laboratories will need to collaborate with doctors and genetic counsellors to ensure that patients receive the best treatment and guidance possible.

Better methods to educate and train healthcare workers about PM must be created and executed if diverse stakeholders are to embrace it.

12.3.3 Social and Ethical Issues

In the practice of PM, the ethics of patient information is a major concern. For instance, various ethical issues arise when a disease is discovered by coincidence.

Another potentially deadly disease may be identified when a patient is being checked for one. This becomes a serious problem if no treatment options are available because the patient's prognosis doesn't quite improve during this time. Some patients might experience some intense emotions to the realization of the new diagnosis and its prognosis, thereby raising further ethical considerations (Brothers and Rothstein, 2012).

False positives, which could occur when genetic variables are misread and a patient is diagnosed with an illness they do not have, are another source of worry. This can lead to several ethical issues. This would include not just adverse emotional consequences but also the adoption of inappropriate therapies that can be damaging to patients and result in excessive healthcare expenses. Another concern is the rights of family members. Some believe that because genetic disorders are hereditary and since early treatment can help prevent the condition from worsening, the family should be informed if they are at risk. Currently, the patient has the option of informing their family or not.

The stigmatization of people with particular diseases is another issue that has to be addressed. Many concerns may arise if a patient's genetic data were released. For example, insurance firms may misuse this knowledge and refuse to give coverage to specific people with genetic diseases. Employers may also use this information to decide whether or not to hire someone based on their genetic condition. As a result, the necessity of genetic information security and confidentiality is emphasized.

To perform at its best, precision medicine requires a large amount of genetic data from a broad category of people. If and when such a massive volume of data is obtained, it is uncertain who will own it. Individuals are unable to access their genetic data, and the government has no control over the information. Apart from legal issues, the collection and storage of such massive volumes of data raise privacy concerns. For this reason, enacting such a policy would most likely be met with fierce opposition.

12.3.4 Cost

PM would minimize redundant efforts, readmissions, and aid in disease prevention, ending the healthcare funding haemorrhage. However, getting to this point necessitates a significant investment in infrastructure for collecting, storing, and exchanging data. To safeguard the data, further investments must be made in security infrastructure, and other costs may prove to be a burden. To cure and prevent sickness, precision medicine employs modern technology and processes to sort and identify the causes of diverse ailments.

12.4 THE FUTURE OF PERSONALIZED MEDICINE

The ongoing change, expansion, and progress in healthcare delivery are fuelled by the fast adoption of technology integration into healthcare. Telehealth, artificial intelligence (AI), nanomedicine, tricorders, robotic surgery, and other innovations are just a few of the many inventions being researched to improve the health of the general populace and healthcare delivery in the future (Vicente et al. 2020). Despite the numerous obstacles that have been and will continue to be encountered in the application of PM, research has demonstrated the importance of PM and its potential to revolutionize clinical medicine and patient care.

AI is becoming an integral aspect of PM. Researchers are combining AI and PM to revolutionize healthcare. As previously stated, knowledge and expertise, as well as the need for guidance to translate these genetic discoveries into therapy, are major challenges in the application of PM. This is where AI comes in, as it has proven to be effective in genomic interpretation and prediction of appropriate therapy. For example, AI has helped to revolutionize the treatment of medulloblastoma in paediatric patients. Following AI-assisted genome analysis, whole-brain radiation was avoided in the treatment of medulloblastoma, and the disease was managed only with chemotherapy (Cesuroglu et al. 2016).

Another way AI is promoting PM is through its use in pharmacology to ensure appropriate drug dosages for each individual and to explain variability in drug responses in different individuals by integrating genetic information with the mechanism of action of diverse medicines. Through disease prevention, early detection, and precise disease management, the integration of AI into PM would ultimately reduce the burden of disease in the general population.

The presence of biobanks has made the future of PM a possibility. Biobanks are facilities that collect, store, and preserve biospecimens or biological information from animals, plants, and people for research purposes in a transparent manner that complies with legal, ethical, and regulatory requirements. Biobanks are closely linked to the development of PM and has played a critical part in its evolution, and continues to do so. They are an important source of biospecimen samples, genetic information, and clinical data. Medical records, genetic data, and population demographics are all available through certain biobanks (Gaziano et al. 2016).

PM research relies on sample analysis and clinical data, and a large number of these are necessary to arrive at more accurate, precise, and timely results and conclusions. The more well-characterized and top-quality samples provided by these biobanks, the faster research will advance, resulting in faster PM adoption and implementation. The systematic collection of high-quality human samples is crucial to the success of future therapies. Biobanks are presently required to standardize tissue collection to improve science quality and genuinely support research (Rogers et al. 2011; Catchpoole, 2017; Malsagova et al. 2020).

It is envisaged that in years to come, with the implementation of PM, there will be a new generation of healthcare providers who are better informed, engaged, skilled and empowered. There would be more focus on digital literacy and a need to be skilled in interpreting genomic information and making more individualized decisions on their findings.

12.5 ROLE OF PUBLIC HEALTH IN MEDICINE

The overall health of a group of people as a population living in a specified geographical location at a particular time is often addressed as public health. It plays a complementary role with personalized healthcare on the health of a population. Personalized healthcare focuses on the biological science of health (Overby & Tarczy-Hornoch, 2013) while public health focuses on social, behavioural, and environmental aspects of health and stresses on prevention of diseases (Shi et al., 2009). Over the years, there has been clinical evidence identifying the huge contribution of public health to preventive medicine (Coughlin, 1999). The focus of care in medicine is slowly drifting away from curative, with more emphasis being placed on prevention which underscores the important part public health plays in medicine. It caters for the health of the community while clinical PM caters for individuals with a disease by individualizing treatment.

PTH has had enormous triumphs in several fields of medicine that cannot be ignored. The reduction in incidence and mortality of some diseases can be attributed to the emergence of PTH (Tulchinsky & Varavikova, 2014). It applies the range of social and community interventions focused on the primary prevention of diseases to preserve and protect health. These interventions include but are not limited to public education, awareness, immunization campaigns, urban design, laws and regulatory functions, and surveillance. All of these interventions work by influencing social, nutritional and behavioural factors to understand health. Public health has been able to effectively pass across to the population the need for sanitation, food safety, improved nutrition and reduction of child and maternal mortality through immunization programs. These have been done through the creation of awareness, continuous public education, and availability of health services at the community level which is fundamental to public health.

PTH is also crucial in the control of infectious diseases and containing disease outbreaks at the community level. During outbreaks, PTH is concerned with tasks like detecting epidemics/pandemics earlier to avoid them becoming widespread, forecasting infection spread rate, and developing functional and potent control/management strategies (Anwar et al., 2020). Public health handles issues that are likely to arise from outbreaks such as fear and worries about the disease among the population, limitation of spread, contact tracing of an infected person, surveillance and reporting of disease. It also plays a huge role in dispelling all myths and false information circulating in the media about an outbreak while also educating the people on the appropriate information as false information about an outbreak can be as disastrous as the outbreak itself. In the past, previous pandemics that have been tackled globally by public health efforts include smallpox, influenza, polio, Ebola, the current COVID-19 (Madhav et al., 2017; Rahimi and Talebi Bezmin Abadi, 2020; Visvikis-Siest et al. 2020).

An aspect of public health is also epidemiology which deals with frequency, distribution of diseases, and analysis of data derived to determine disease control. Public health is also important in research by providing scientific evidence about diseases. More light has been shed on symptomatology and the treatment of diseases through high-level research works. Additionally, when it comes to the management and control of non-communicable diseases which includes autoimmune diseases, hypertension, Parkinson's disease heart diseases and diabetes, PTH plays a crucial role. When it comes to effective ways of reducing the prevalence of these diseases, targeting people at the community level has proven to be a better option and more useful than the individual approach which takes a longer time (Talic et al., 2021). PTH also played an important part in occupational health by ensuring that employees observe standard precautions in limiting the risk of disease among workers (Davis & Souza, 2009).

Practices at this level include regular screening, monitoring of hazard levels, and continuous reduction of exposure to hazards. These are issues that public health has helped in tackling. Diseases usually develop at the community level before the individual presentation at hospitals. Hence, if the risks to diseases are tackled at the community level, exposure of individuals is reduced, and consequently, reduce prevalence and need for treatment at hospital individually. Public health physicians have largely used this approach in practising preventive and social medicine.

12.6 PUBLIC HEALTH (PTH) SYSTEMS

The definition of CDC (2020), states that PTH systems refer to every organization either private or public that provides crucial or necessary health services to a given population. The PTH system is usually a combined effort between government health agencies and private organizations and can run hierarchically from local to state to federal levels. Public health delivery systems exist as complex and adaptive systems that operate through the interactions of multiple players that include businesses, educational institutions, community leaders, healthcare providers, and public health agencies among others (Mays et al., 2006). They are all important in providing an effective healthcare service delivery. The legislation and regulatory role rest on the government (Beaglehole et al., 2004).

The effectiveness of PTH systems can vary widely depending on the availability of resources and personnel. Several authors have tried to understand the effect and roles several local PTH systems plays and how they affect the accessibility and success rate of public health services including ways of improving them. Theories of organizational sociology and industrial organization have been used in explaining the public health system by suggesting how activities executed can structure the availability, accessibility and organization of resources in the communities to be served (Handler et al., 2001; Salancik & Pfeffer, 1978). An example to illustrate this is the outbreak of a disease which becomes the primary responsibility of the public health system to curtail this disease. In executing its role in curtailing, the ability depends on the resources available which include knowledge, capital, and enough cases to be able to identify clinical and molecular characteristics of such diseases. Additionally, the community also has to be receptive to solutions in curtailing the disease to make the public health system effective. There has to be an organized and follow-up protocol in ensuring the effectiveness of the system. When the targeted community refuses to follow up on laid-down protocols, it waters down the efficacy of the PTH systems in complementing personalized healthcare, alongside the burden of overall healthcare shifts mainly to personalized healthcare.

There is not so much evidence in the literature of research on the public health system characteristics and challenges associated with it. However, Mays et al. (2006) noted how the majority of finance for most local government public health agencies comes from local governments and state governments which makes it susceptible to negative effects of political instability. Political stability has been recognized as a factor that affects public health systems (Ranabhat et al., 2020). A new government may abandon existing projects that large resources have been spent on leading to wastage and mismanagement of resources. Similarly, most public health agencies are challenged with the problem of staffing. Neglected aspects of the public health system also include leadership problems, failure of advocacies, and legalization of certain health issues which impair the quality of healthcare provided to the population. Furthermore, the availability of resources and structures in respective organizations determines the activities of PTH systems (Mays et al., 2004). Generally, vast PTH systems are greatly influenced its activities and this dictates the population size of people that benefit from it. Large PTH systems carry out their activities on a large scale, this includes health education, community enlightenment through campaigns, and disease surveillance with the target populace as the beneficiaries. This allows the spread of cost across taxpayers with a larger pool of financial income (Mays et al., 2001). A larger public health system has been noted to have a more effective performance compared to smaller ones in activities that are core to the public health system (Mays et al., 2004).

It is noteworthy how the public health system responds to surges in the number of diseases and data from personalized healthcare, especially for rare diseases or common diseases with increasing incidence. It then goes on to use epidemiology to develop interventions to respond to the surges in the number of disease cases. In communities where resources are inadequate, public health takes on the role of personalized healthcare by diverting resources to care for members of the community while leaving fewer resources for the core services of public health. There is a diversion of resources to health promotion, diagnosis and treatment of individuals rather than the overall safety of the community due to inadequate healthcare resources. Adequate government funding is an essential ingredient in facilitating healthcare of public health. The direction for a new era of public health system requires a government with dedication and strong will, responsive leadership, provision of infrastructure and adequate resources for personnel complemented by public health research, using the available tools maximally (Beaglehole et al., 2004).

12.7 CONVERGENCE OF PUBLIC HEALTH AND PERSONALIZED MEDICINE

With the emergence of precision medicine in ensuring a more focused and individualized approach in treating diseases, there is a need for the internalization of PM into the public health system to ensure efficiency, effectiveness and sustainability. The challenges that are faced with both systems have been highlighted above and common problems shared include funds, leadership, sustainability, political instability, and policy, among others (Sedda et al. 2019). Although treatment is to be tailored to individuals, the concept of PM can be extended to subpopulations with identifiable genomics, biological, and environmental characteristics which translates to a new concept of precision public health. This could involve classifying people into subpopulations based on similar features like behaviours and genetic information to aim at better intervention and treatment outcomes (Velmovitsky et al., 2021).

Precision public health is an emerging concept that has been discussed by frontiers in the field of public health and it involves disease prevention, health promotion and reduction in health indices disparities among different populations through the utilization of new methods and technologies which helps in tailoring interventions (Khoury, 2019).

Some of these emerging interventions include electronic health records, genomic sequencing, and patient-generated health data. The collection of patient-generated data provides specific information to physicians such as levels of physical activity, lifestyle behaviours, socioeconomic characteristics, and also eases the use of such data for public health purposes (Vicente et al. 2020). This integration of PM into a public health system also helps in the judicious use of funds by improving output.

Without the knowledge of the population's health status, it might be difficult to hypothesize an individual's health problems as public health data can provide insight into what personal risk factors an individual should be evaluated for in PM. Similarly, authors have opined how the collection of several individual genomes through genomic sequencing in precision medicine can help in the recognition of variance and patterns across genes of different populations, which can as well impact scientific knowledge on therapeutic options (Ramaswami et al., 2018). All of these further cement the idea of complementary efforts between PM and public health and how the two separate fields can be integrated into precision public health. Hence, the need to marry the two concepts together into providing precise medicine.

12.8 CONCLUSION

PM is unquestionably the future of healthcare, as the industry shifts to a more tailored approach. It's not a new concept; it's been around for quite some time. Technology advancements, on the other hand, have continuously opened up new frontiers and opportunities for PM. Because everyone is unique, each illness should be treated as such. This would lead to more effective methods of disease prevention, diagnosis, and treatment. Because people will always live in close quarters, public health will always be important, and some health problems will always be specific to a particular group of people. PM and public health are two distinct concepts of intervention in healthcare and they have both been described to be very beneficial to healthcare as a whole. Exploration of emerging technologies in the fields of PM and public health has progressively advanced healthcare. However, both systems are interdependent and a comparison of both concepts shows how none is preferable to the other as they are both interwoven and faced with challenges that can be solved via a convergent model of the two concepts. Unlike PM, public health would be expected to respond to circumstances such as disease outbreaks and community healthcare as quickly as possible and institute preventive measures. This reinforces a need for convergence for improved outcomes. However, the majority of the studies on this topical issue have been qualitative with very scanty quantitative evidence. There is a need for future research to provide quantitative evidence as regards the benefits and the possible challenges to a convergent approach of PM and public health.

REFERENCES

Agyeman, A. A., and Ofori-Asenso, R. (2015). Perspective: Does personalized medicine hold the future for medicine. *Journal of pharmacy & bioallied sciences*, 7(3), 239.

Anwar, A., Malik, M., Raees, V., and Anwar, A. (2020). Role of mass media and public health communications in the COVID-19 pandemic. *Cureus*, 12(9), e10453. https://doi.org/10.7759/cureus.10453

Appel, L. J. (2003). Lifestyle modification as a means to prevent and treat high blood pressure. *Journal of the American Society of Nephrology*, 14(suppl 2), S99–S102.

Beaglehole, R., Bonita, R., Horton, R., Adams, O., and McKee, M. (2004). Public health in the new era: Improving health through collective action. *The Lancet*, 363(9426), 2084–2086. https://doi.org/10.1016/S0140-6736(04)16461-1

Brothers, K. B., and Rothstein, M. A. (2012). Ethical, legal, and social implications of incorporating personalized medicine into. *Journal of Law, Medicine & Ethics*, 40(2), 394–400.

Catchpoole, D. R. (2017). Getting the message about biobanking: Returning to the basics. *Journal of Biorepository Science for Applied Medicine*, 5, 9–21.

CDC. (2020, September 8). Original essential public health services framework—OSTLTS. https://www.cdc.gov/publichealthgateway/publichealthservices/originalessentialhealthservices.html

Cesuroglu, T., Syurina, E., Feron, F., and Krumeich, A. (2016). Another side of the coin for personalised medicine and healthcare: A content analysis of 'personalised' practices in the literature. *BMJ Open*, 6(7), e010243.

Coughlin, S. S. (1999). The intersection of genetics, public health, and preventive medicine. *American Journal of Preventive Medicine*, 16(2), 89–90. https://doi.org/10.1016/S0749-3797(98)00135-4

Davis, L., and Souza, K. (2009). Integrating Occupational health with mainstream public health in Massachusetts: An approach to intervention. *Public Health Reports*, 124(Suppl 1), 5–15.

Drijver, M., and Woudenberg, F. (1999). Cluster management and the role of concerned communities and the media. *European Journal of Epidemiology*, 15(9), 863–869.

Gaziano, J. M., Concato, J., Breeding, M., Fiore, L., Pyarajan, S., Breeling, J., Whitbourne, S., Deen, J., Shannon, C., Humphries, D., Guarino, P., Aslan, M., Anderson, D., LaFleur, Rene, Hammond, T., Schaa, K., Moser, J., Huang, G., Muralidhar, S., Przygodzki, R., and O'Leary, T. J. (2016). Million veteran program: A mega-biobank to study genetic influences on health and disease. *Journal of Clinical Epidemiology*, 70, 214–223. https://doi.org/10.1016/j.jclinepi.2015.09.016

Handler, A., Issel, M., and Turnock, B. (2001). A conceptual framework to measure performance of the public health system. *American Journal of Public Health*, 91(8), 1235–1239. https://doi.org/10.2105/AJPH.91.8.1235

Hood, L., and Flores, M. (2012). A personal view on systems medicine and the emergence of proactive P4 medicine: Predictive, preventive, personalized and participatory. *New Biotechnology*, 29(6), 613–624.

Horgan, D., Romao, M., and Hastings, R. (2017). Pulling the strands together: MEGA steps to drive European genomics and personalised medicine. *Biomedicine Hub*, 2(Suppl. 1), 1–11.

Israel, B. A., Schulz, A. J., Parker, E. A., and Becker, A. B. (1998). Review of community-based research: Assessing partnership approaches to improve public health. *Annual Review of Public Health*, *19*(1), 173–202.

Karthika, S., and Ragunath, P. K. (2007) Pharmacogenomics – A new perspective in clinical healthcare. *Sri Ramachandra Journal of Medicine*, 1(2), 34–40.

Khoury, M. (2019). *Precision public Health and Precision Medicine: Two Peas in a Pod*. https://scholar.google.com/scholar_lookup?title=Precision+Public+Health+and+Precision+Medicine:+Two+Peas+in+a+Pod&author=MJ+Khoury&

Kreuter, M. W., Farrell, D. W., Olevitch, L. R., and Brennan, L. K. (2013). *Tailoring health messages: Customizing communication with computer technology*. Routledge.

Kubota, L. T., Da Silva, J. A. F., Sena, M. M., & Alves, W. A. (Eds.). (2022). *Tools and Trends in Bioanalytical Chemistry*. Berlin/Heidelberg, Germany: Springer.

Madhav, N., Oppenheim, B., Gallivan, M., Mulembakani, P., Rubin, E., and Wolfe, N. (2017). Pandemics: Risks, impacts, and mitigation. In D. T. Jamison, H. Gelband, S. Horton, P. Jha, R. Laxminarayan, C. N. Mock, and R. Nugent (Eds.), *Disease control priorities: Improving health and reducing poverty* (3rd ed.). The International Bank for Reconstruction and Development/The World Bank. http://www.ncbi.nlm.nih.gov/books/NBK525302/

Malsagova, K., Kopylov, A., Stepanov, A., Butkova, T., Sinitsyna, A., Izotov, A., and Kaysheva, A. (2020). Biobanks—A platform for scientific and biomedical research. *Diagnostics*, 10(7), 485.

Marcon, A. R., Bieber, M., and Caulfield, T. (2018). Representing a "revolution": How the popular press has portrayed personalized medicine. *Genetics in Medicine*, 20(9), 950–956.

Marson, F. A., Bertuzzo, C. S., and Ribeiro, J. D. (2017). Personalized or precision medicine? The example of cystic fibrosis. *Frontiers in Pharmacology*, 8, 390.

Mathur, S., and Sutton, J. (2017). Personalized medicine could transform healthcare. *Biomedical Reports*, 7(1), 3–5.

Mays, G. P., Halverson, P. K., Baker, E. L., Stevens, R., and Vann, J. J. (2004). Availability and perceived effectiveness of public health activities in the nation's most populous communities. *American Journal of Public Health*, 94(6), 1019–1026.

Mays, G. P., Halverson, P. K., and Stevens, R. (2001). The contributions of managed care plans to public health practice: Evidence from the nation's largest local health departments. *Public Health Reports*, 116(Suppl 1), 50–67.

Mays, G. P., McHugh, M. C., Shim, K., Perry, N., Lenaway, D., Halverson, P. K., and Moonesinghe, R. (2006). Institutional and economic determinants of public health system performance. *American Journal of Public Health*, 96(3), 523–531. https://doi.org/10.2105/AJPH.2005.064253

Noell, G., Faner, R., and Agustí, A. (2018). From systems biology to P4 medicine: Applications in respiratory medicine. *European Respiratory Review*, *27*(147), 170110. https://doi.org/10.1183/16000617.0110-2017

Overby, C. L., and Tarczy-Hornoch, P. (2013). Personalized medicine: Challenges and opportunities for translational bioinformatics. *Personalized Medicine*, 10(5), 453–462.

Rahimi, F., and Talebi Bezmin Abadi, A. (2020). Tackling the COVID-19 pandemic. *Archives of Medical Research*, 51(5), 468–470. https://doi.org/10.1016/j.arcmed.2020.04.012

Ramaswami, R., Bayer, R., and Galea, S. (2018). Precision medicine from a public health perspective. *Annual Review of Public Health*, 39, 153–168.

Ranabhat, C. L., Jakovljevic, M., Dhimal, M., and Kim, C.-B. (2020). Structural factors responsible for universal health coverage in low- and middle-income countries: Results from 118 countries. *Frontiers in Public Health*, 7. https://www.frontiersin.org/article/10.3389/fpubh.2019.00414

Rogers, J., Carolin, T., Vaught, J., and Compton, C. (2011). Biobankonomics: A taxonomy for evaluating the economic benefits of standardized centralized human biobanking for translational research. *Journal of the National Cancer Institute Monographs*, 2011(42), 32–38.

Sadkovsky, I.A., Golubnitschaja, O., Mandrik, M. A., Studneva, M. A., Abe, H., Schroeder, H., Antonova, E. N., Betsou, F., Bodrova, T. A., Payne, K., and Suchkov, S. V. (2014). PPPM (Predictive, Preventive and Personalized Medicine) is a new model of the national and international healthcare services and thus a promising strategy to prevent disease: from basics to practice. *International Journal of Clinical Medicine*, *5*(14), 855–870.

Salancik, G. R., and Pfeffer, J. (1978). *The external control of organizations: A resource dependence perspective*. Harper & Row.

Sedda, G., Gasparri, R., and Spaggiari, L. (2019). Challenges and innovations in personalized medicine care. *Future Oncology*, 15(29), 3305–3308.

Shi, L., Tsai, J., and Kao, S. (2009). Public health, social determinants of health, and public policy. *J Med Sci*, 29(2), 43–59.

Simmons, L. A., Dinan, M. A., Robinson, T. J., and Snyderman, R. (2012). Personalized medicine is more than genomic medicine: Confusion over terminology impedes progress towards personalized healthcare. *Personalized Medicine*, 9(1), 85–91.

Talic, S., Shah, S., Wild, H., Gasevic, D., Maharaj, A., Ademi, Z., Li, X., Xu, W., Mesa-Eguiagaray, I., Rostron, J., Theodoratou, E., Zhang, X., Motee, A., Liew, D., and Ilic, D. (2021). Effectiveness of public health measures in reducing the incidence of covid-19, SARS-CoV-2 transmission, and covid-19 mortality: Systematic review and meta-analysis. *BMJ*, 375, e068302. https://doi.org/10.1136/bmj-2021-068302

Thompson, S., Kohli, R., Jones, C., Lovejoy, N., McGraves-Lloyd, K., and Finison, K. (2015). Evaluating health care delivery reform initiatives in the face of "cost disease". *Population Health Management*, 18(1), 6–14.

Travensolo, R. F. D., Ferreira, V. G., Federici, M. T., Lemos, E. G. M. D., and Carrilho, E. (2022). Microarrays application in life sciences: The beginning of the revolution. In Lauro Tatsuo Kubota, José Alberto Fracassi da Silva, Marcelo Martins Sena, Wendel Andrade Alves (Eds.) *Tools and trends in bioanalytical chemistry* (pp. 483–496). Springer.

Tulchinsky, T. H., and Varavikova, E. A. (2014). Chapter 1—A history of public health. In T. H. Tulchinsky and E. A. Varavikova (Eds.), *The new public health* (3rd ed., pp. 1–42). Academic Press. https://doi.org/10.1016/B978-0-12-415766-8.00001-X

Velmovitsky, P. E., Bevilacqua, T., Alencar, P., Cowan, D., and Morita, P. P. (2021). Convergence of precision medicine and public health into precision public health: Toward a big data perspective. *Frontiers in Public Health*, 9, 561873. https://doi.org/10.3389/fpubh.2021.561873

Vicente, A. M., Ballensiefen, W., and Jönsson, J. I. (2020). How personalised medicine will transform healthcare by 2030: The ICPerMed vision. *Journal of Translational Medicine*, 18(1), 1–4.

Visvikis-Siest, S., Theodoridou, D., Kontoe, M. S., Kumar, S., and Marschler, M. (2020). Milestones in personalized medicine: From the Ancient Time to Nowadays—the Provocation of COVID-19. *Frontiers in Genetics*, *11*, 569175. https://doi.org/10.3389/fgene.2020.569175

13 IoT-Driven and the Role of Robotic Technology in Healthcare

Akinola Samson Olayinka
Edo State University, Uzairue, Nigeria

Tosin Comfort Olayinka
Wellspring University, Benin City, Nigeria

Charles Oluwaseun Adetunji
Edo State University, Uzairue, Nigeria

Clement Atachegbe Onate
Landmark University, Omu-Aran, Nigeria

Olusanmi Ebenezer Odeyemi
Federal College of Animal Health and Production Technology, Ibadan, Nigeria

Onyijen Ojei Harrison
Samuel Adegboyega University, Ogwa, Nigeria

Oluwakemi Mary Odeyemi
Joseph Ayo Babalola University, Ikeji Arakeji, Nigeria

Oluwafemi Adebayo Oyewole
Federal University of Technology, Minna, Nigeria

K. I. T. Eniola
Joseph Ayo Babalola University, Ikeji Arakeji, Nigeria

DOI: 10.1201/9781003309468-13

INTRODUCTION

Development in modern robotics is traceable to the invention of universal automation (Unimate), a reprogrammable manipulator by George C. Devol in the 1950s (Stanford 2022). Unimate was later acquired by Joseph Engleberger and was modified into an industrial robot. Between 1966 and 1972, the first artificial intelligence (AI) based mobile robot, known as "Shakey" was developed by Charles Rosen working at the Artificial Intelligence Center of the Stanford Research Institute (STI) with his team. Robotics application was first introduced in the medical field in 1985 with the utilization of a robotic arm to support the stereotactic biopsy in neurosurgery. That was the first time robots were employed in surgery (Beasley 2012; Petrescu 2019).

Recent advances in technology and innovation in the medical field and robotics are revolutionizing healthcare delivery in surgery, surgical-assistance, dispensary, caregiving, receptionists, and treatment of patients leading to a reduction in the time doctors and medical personnel spend with patients (Beasley 2012; Chen *et al.* 2016; Nguyen *et al.* 2019; Johanson *et al.* 2020; Tavakoli *et al.* 2020; Ginoya *et al.* 2021; Sinha *et al.* 2021). Developments in the field of AI, machine learning, human computer interface (HCI), computer vision, human computer interface (HCI) and data analytics have changed the narratives in the health sector with the introduction of robots to healthcare delivery.

Developments in robotic technology have seen robots being put into action in hospital to help medical workers with the hope of ensuring better quality healthcare delivery. During the COVID-19 outbreak, in order to decrease infection risk, medical facilities started using robots for a wider range of jobs. (Tavakoli *et al.* 2020; Van den Eynde *et al.* 2020; Zeng *et al.* 2020; Wang *et al.* 2021). It's become clear that health robots' cost-cutting and risk-reduction capabilities benefit a wide range of industries. Infectious health units, for instance, can deploy robots to clean and prepare patient rooms on their own, eliminating interaction with people and encouraging the "high-touch to high-tech" notion (Zeng *et al.* 2020). In hospitals, the time required to discover, identify, and administer medications to patients is reduced by robots with AI-enabled medicine identifier software.

Petrescu identified various robotic services in medical healthcare delivery to include a medical procedure that requires high precision where the size is small and any human error can lead to fatality (Petrescu 2019). Robots help doctors in operations on the heart, brain, and kidneys, as well as bone implants and the repair of broken bones, cartilage, and muscles.

Medical activities are premised on data from diverse sources like patient personal data such as vital signs and photographs of body tissues and organs as well as general medical knowledge, and medical experiences. Robots can be very useful in X-ray related activities without jeopardizing the health of the medical team.

IBM developed Robo-doc, the first robot used in therapeutic therapy, in the late 1980s. Transurethral prostate excision was the first time a robot was used in human surgery. In 1993, Computer Motion, Inc. introduced the Automated Endoscopic System for Optimal Positioning (AESOPTM), a voice-controlled arm used to aid tools and optics in laparoscopic surgery. The Food and Drug Administration in the US has recognized the AESOPTM 2000 version as the first human-controlled robot.

In 1998, the ZEUS Microsurgical Robotic System was introduced in Germany by Reichenspurner. Currently, the DaVinci system is the most complex and effective robot in use (Mettler *et al.* 1998; Petrescu 2019; Oniani *et al.* 2021).

With the introduction of laparoscopy and computer technology, the field of surgery has entered a new age. Surgical robots are being developed to improve the efficacy of surgical medical treatments. Patient-specific data such as vital signs, photographs of human tissues and organs, medical information in general as well as medical experiences are used to guide medical activities. A robot is usually far more precise than a human. This is the driving force for the use of CAD/CAM technologies. Robots can be used successfully without risking the medical team's health if the patient has to be exposed to an X-ray.

Johanson *et al.* studied the effect of a smiling receptionist robot on the perception of hospital patients. They projected the need to create and investigate robots' behavioural capacities so that individuals can have appropriate and comfortable intuition with them. The use of social communication techniques such as smiling and addressing someone by their first name as obtainable during human interactions is presented as a way of improving patients' perception of robot personality. The study concluded that a smiling robot can attract reciprocate smiling from users (Johanson *et al.* 2020).

IoT COMMUNICATION PROTOCOL

IoT hardware covers a variety of devices such as routing devices, bridges, switches, sensors, controllers, microcontroller modules, and so on. Hardware and software component of the IoT are the building block of the IoT devices. Dedicated hardware devices are used in implementing the interaction with the real environment. Software that analyzes inputs and operates the system is run on microcontrollers.

Internet of things (IoT) information communication technology (ICT) is a game-changer in human–human interaction, human–things interaction, and things–things interaction and information exchange. Smart devices can now connect, move, and make decisions for people; this technology is referred to as "connectivity for anything." The IoT ecosystem consists of a collection of smart gadgets that can connect with each other at any time and from any location. However, limitations such as storage volume, radio range, and power are impeding IoT growth, necessitating the development of a communication protocol to address these issues.

These IoT devices provide a variety of duties and services, such as activation of the system, security, action requirements, communication protocols, and identification of specific support objectives and actions.

The IoT technology system is nothing if the proper IoT protocols are not in place. IoT technology requires the right IoT protocol for it to function as expected. IoT devices would be considered worthless in the absence of IoT protocols and associated standards. IoT protocols and standards allow hardware to communicate and interact with each other as needed. When the concept of the IoT comes to the fore, protocols and standards are sometimes forgotten. Most of the time, the attention is focused on communication. Gadgets, smart sensors, gateways, servers, and user applications must all be connected for the IoT to work; however, without the right

IoT protocols, communication would be nearly impossible. IoT communication protocol is subdivided into data protocols and network protocols for data exchange applications and network communications respectively. Figures 13.1 and 13.2 show IoT network protocol and IoT data protocols respectively. Some of the popular data and network protocols are discussed below.

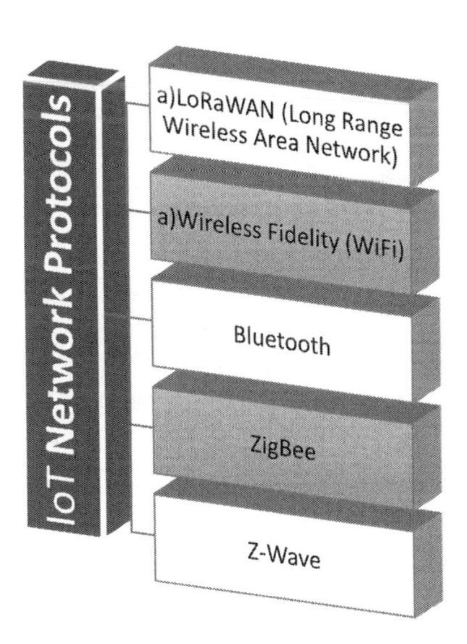

FIGURE 13.1 IoT network protocols.

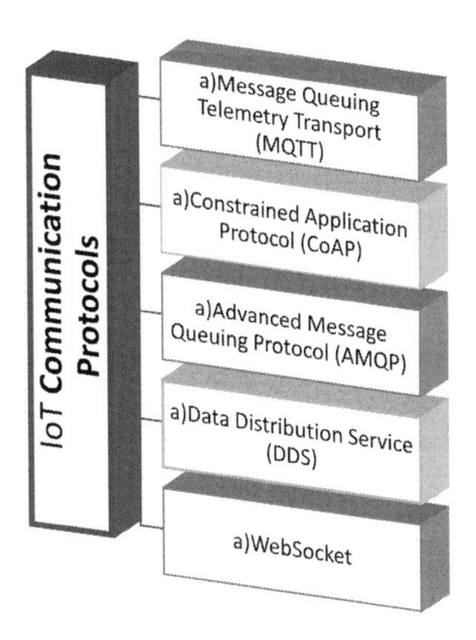

FIGURE 13.2 IoT data protocols.

DATA PROTOCOLS FOR IoT

Message Queuing Telemetry Transport

Message Queuing Telemetry Transport (MQTT) is a straightforward data protocol. It is a messaging protocol system that allows data to be easily transferred between devices. The design of MQTT is its key selling point. It has a basic and lightweight physical build, allowing it to create energy-efficient technologies. It uses the TCP/IP protocol as well. Data protocols for the IoT were created to deal with unpredictable communication networks. This has become a demand in the IoT world due to the surge in the number of small, low-cost, and low-power gadgets that have appeared in the network during the last few years. Despite its extensive acceptance – most notably as an IoT standard with industrial applications – MQTT does not have a standardized data representation or device administration mode. As a result, platform or vendor-specific data and device management capabilities must be developed.

Constrained Application Protocol

The Constrained Application Protocol (CoAP) is a unique internet application layer protocol for restricted or constrained devices with detailed specifications in RFC 7252 (Shelby *et al.* 2014). The protocol is tailored to meet the requirements of HyperText Transfer Protocol (HTTP) based IoT systems. The CoAP is a web transfer protocol designed for usage with constrained nodes and applications where low power is an important factor for consideration. The nodes are often constrained to 8 bit while microcontrollers are constrained with small quantities of memory (RAM and ROM). CoAP is targeted at applications in machine-to-machine (M2M) communications. The protocol allows “nodes,” or constrained devices, to connect with the rest of the internet by adopting compatible protocols. CoAP is intended for application in devices on a constrained network (Shelby *et al.* 2014). Although the internet service infrastructure is accessible to all IoT devices and can be used by them, many implementations find it to be excessively bulky and energy consuming. As a result, many in the IoT industry believe HTTP is unsuited for IoT. CoAP circumvents this limitation by repurposing the HTTP framework for use in limited devices and network environments. It has very low overheads, is easy to use, and can support multicast broadcasting. As a consequence, it’s ideal for devices with minimal resources, such as IoT embedded systems or Sensor nodes.

Advanced Message Queuing Protocol

The Advanced Message Queuing Protocol (AMQP) is a basic application layer protocol that supports a variety of messaging apps and methods of communication. AMQP is an open standard that was created in 2003 by John O’Hara as an application layer protocol that allows servers to send and receive transactional messages (Vinoski 2006; O’Hara 2007). The primary work of AMQP include handling messages in terms of receiving, sending, queueing, storing, and creating connection for various elements to communicate. It’s most typically used in settings that require server-based analytical systems, like the financial sector, for its high security and reliability. AMQP is not suited for IoT sensor devices with limited memory due to its weight. Consequently, its use in the IoT is currently somewhat restricted.

Data Distribution Service

Data Distribution Service (DDS) is yet another extensible IoT protocol that provides increased IoT connectivity. DDS operates on the same publisher–subscriber mechanism as MQTT. It can be used in a wide range of settings, including the cloud and very small devices. For this, it's perfect for real-time and embedded systems. Furthermore, the DDS protocol enables compatible data exchange regardless of hardware or software platform. It is often considered as the first open worldwide middleware IoT standard.

WebSocket

WebSocket is a technology that allows users to communicate in full duplex over a single TCP connection. In 2011, the Internet Engineering Task Force (IETF) published RFC 6455, which defined the WebSocket protocol. It is a TCP-based protocol that is self-contained (Shiraishi 2013; El Ouadghiri *et al.* 2020). The World Wide Web Consortium known as W3C created the WebSocket application programming interface (API) for use in web application. This specification defines APIs that enable internet applications to communicate with server-side activities in a bidirectional manner (WebSockets Standard 2022). A solitary communication protocol can carry all messages between the client and the server. WebSocket's communication protocol, like CoAP's, aids in the administration of internet connectivity and data exchange removing many of the difficulties and challenges. It can be used in an IoT network where data is continuously transmitted between various devices. As a result, it's most typically seen in areas that serve as clients or servers. This contains libraries and runtime environments.

NETWORK PROTOCOLS FOR IoT

The IoT, network protocols are used to connect devices via a network. These protocols are often used on the internet. The following are some examples of IoT network protocols.

Long Range Wireless Area Network (LoRaWAN)

LoRaWAN is an open protocol that enables IoT devices to connect over the internet. The lower physical layer is defined by LoRa, while the higher networking layers are missing. LoRaWAN is part of various protocols designed to describe the network's higher levels. LoRaWAN is a protocol in the medium access control (MAC) layer that runs in the cloud. LoRa is a specialized wireless spectrum modulation technology. LoRaWAN has many of the same advantages as 5G, including great range, cost-effectiveness, and high battery efficiency. LoRaWAN's popularity in the IoT market is fast growing. Recent trends have led industry experts to predict that the technology will capture roughly 75% of the IoT market (now valued at over $465 billion), displacing 5G. The long-range communication protocol addresses some essential requirements that 5G networks have yet to address. The understanding of the basic principles two technologies is critical to the comparison of 5G vs. LoRaWAN (Makarchuk 2022).

Wireless Fidelity

Wireless fidelity (Wi-Fi) is the most popular IoT protocol that is well known by most users. A device such as mobile phones, computers or routers can be used to set up a Wi-Fi network to provide access to IoT devices. Wi-Fi links adjacent devices to the internet within a specific range. Another option to use Wi-Fi is to create a Wi-Fi hotspot. By transmitting a signal, mobile devices, phones and computers can connect to other devices through a wireless or wired network. Wi-Fi uses radio waves to transport data at specific frequencies like 2.4 GHz and 5 GHz. In addition, both of these frequency ranges contain a wide range of channels that can be used by a number of wireless devices. As a result, wireless networks are not overwhelmed.

Bluetooth

Bluetooth is a wireless short-range (100 m) technology standard for data communication between fixed or mobile gadgets over short distances. Bluetooth uses the frequency range of 2.40 to 2.48 GHz in the ultra-high frequency (UHF) spectrum of electromagnetic waves. Compared to other IoT network protocols, Bluetooth has a shorter range and tends to frequency hop. Its integration with modern mobile devices, such as smartphones and tablets, as well as wearable technology, such as wireless headphones and automotive music players, has made it popular. The current Bluetooth 4.0 standard includes 40 channels and a 2 MHz bandwidth, allowing for data transfer rates of up to 3 Mb/s. Bluetooth Low Energy (BLE) is a low-power wireless technology that has been created for short-range control and monitoring applications and is projected to be used by billions of devices in the decades to come. This innovative approach has the potential to change the narratives in the IoT applications that require a high level of flexibility, scalability, and low power consumption (Gomez *et al.* 2012; Ghori *et al.* 2020; Pušnik *et al.* 2020).

ZigBee

In the area of IoT, ZigBee-based networks are similar to Bluetooth except that they can transmit data up to a distance of 200 m. It is already popular in IoT applications and significantly outperforms the more widely used Bluetooth. It uses less power and is more secure (Fan *et al.* 2017). It's a simple packet data transfer protocol that's common in low-power devices like microcontrollers and smart sensors. It also scales well to tens of thousands of nodes. Many IoT hardware experts are responding by producing products that embrace ZigBee's open standard self-assembly and self-healing grid topology architecture. ZigBee networks have been extensively used in various IoT uses like smart monitoring of weather and environmental variable like temperature and humidity (Hussein *et al.* 2020; Xiao & Li 2020); medical application of blood pressure monitoring (Adi & Kitagawa 2019); and unmanned aerial vehicle (UAV) applications (Pereira *et al.* 2020)

Z-WAVE

The Z-Wave is another wireless protocol and gaining popularity gradually. It's a radio frequency (RF) signal based low-power communication protocol. It uses the 800–900 MHz radio frequency to operate. It's largely utilized for IoT household applications for control, monitoring, and status reading in residential. It provides entire mesh networks without the requirement for a coordinator node, and because it runs in the sub-1GHz band, it is immune to interference from Wi-Fi and other 2.4 GHz wireless technologies such as Bluetooth and ZigBee. ITU-T Recommendation G.9959 defines the Z-Wave PHY and MAC layers (Badenhop *et al.* 2017; Z-Wave Alliance 2021).

WEARABLE IoT DEVICES

Wearable IoT devices are a class of electronic devices that utilize the technology of the IoT which can be worn on the body as accessories, implanted in user clothing or the body, or even used as decorative devices on different parts of the human body for different health or scientific purposes. Most of these technologies seem to onlookers or non-users to be a form of beautification but there are a lot of roles that such technology could perform. A lot of wearable IoT technologies have realistic applications. They could be physically put on the body or implanted in human bodies. They make lives easier. Wearable IoT devices aren't necessarily transformative or some out-of-the-box phenomenon. Wearable IoT devices have been widely employed in a variety of applications, such as predicting user intentions at work (Yildirim & Ali-Eldin 2019); hazards monitoring and reduction in the work environment (Zradziński *et al.* 2020), energy management in the biomedical application of tracking human activities, detection of gestures, and monitoring of healthcare (Park *et al.* 2020). Figure 13.3 shows wearable IoT devices on a physically challenged individual restricted to an IoT-enabled wheelchair. IoT wearable devices can include head-worn or wrist-worn straps, smart shirts, smart shoes, smart posture timers, and smart glass. These wearables are powered by sensors that are capable of measuring various data like heartbeat rate, vital signs, sugar levels, GPS location, and posture among others. These devices also provide feedback where necessary to help users adjust to a healthy lifestyle. OMsignal Smart Clothing is a practical example of a wearable IoT for smart healthcare application where various IoT devices are deployed for non-invasive glucose monitoring to prevent hypoglycemia and hyperglycemia, as well as vital sign monitoring for optimal health (OM Signal Inc. 2015; Steinberg *et al.* 2019).

Advantages of wearable IoT devices include the revolution in the human way of life with hands-free and portable devices for personalized healthcare monitoring and control as well as the ability to track our fitness levels and track our whereabouts with GPS.

A major setback to the wearables IoT devices is that the battery life is usually quite short and the hassle of regularly charging them can be burdensome. However, several

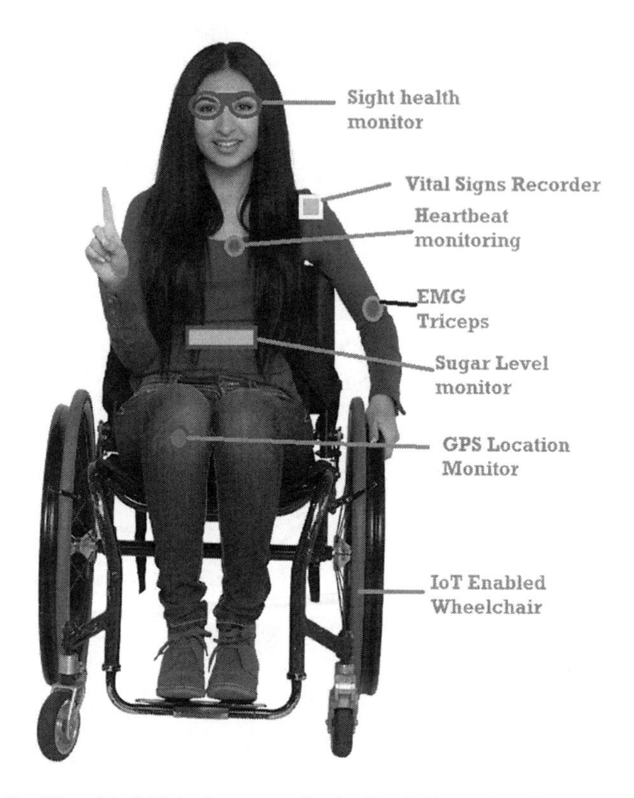

FIGURE 13.3 Wearable IoT devices on a physically challenged individual.

developers are investigating the idea of wireless charging alternatives that would eliminate the need to remove the device from the charger. Inaccurate data measurement is another setback in wearable IoT devices and this can be devasting if data concern has to do with healthcare provision. The future of IoT wearable devices holds promises that it will become commonplace in various applications in healthcare and delivery and industrial workplace monitoring, and smart farming operations (Zhang *et al.* 2021).

RECENT APPLICATION OF IoT AND ROBOTIC TECHNOLOGY IN HEALTHCARE DELIVERY

IoT is becoming more important in robotics technology. The internet of robotic things (IoRT) is being created by bringing together the IoT and robotics groups. The IoT is a network of connected things that includes IoT devices and IoT-enabled physical assets ranging from consumer electronics to sensor-equipped connected technology.

The WWS (Wearable Wellness System) is an all-in-one wearable system that is configured to monitor a group of physiological indicators continuously while moving. The WWS system's underwear is comparable to regular underwear made up of antibacterial and antiallergic ingredients-based yarn to guarantee safe and long-lasting use. The sensors are made of fibres that are immediately woven into the garment during the manufacturing process, allowing them to be incorporated into the garment without any gaps. The connections between sensors are also made with conductive fibres, resulting in a sensorized garment that is completely unobtrusive and extremely comfortable. The shirts are available for both males and females, and a variety of sizes to accommodate a wide range of consumers. A band is also provided for individuals who need to put on and take off the garment fast (WWS – Garment 2022).

Wearable health devices (WHDs) are constantly assisting people with better screening and monitoring of their health and activity levels, with the possibility for earlier diagnosis and treatment. The breakthrough in the shrinking of electronic devices allows for the creation of more durable and flexible wearable devices, contributing to a shift in the method to determine prosperity. Wearable innovation refers to information technology (IT) enabled gadgets that can be worn on the client's body, such as on the wrist, arm, or head, as well as recent logical advancements on the territory (electrocardiogram, pulse, circulatory strain, breath rate, blood oxygen immersion, blood glucose, skin sweat, capnography, internal heat level, movement assessment, cardiovascular implantable gadgets, and encompassing boundaries). The primary ideas underpinning the creative developments in healthcare are wearable gadgets and networks between these gadgets and PCs.

These capacities include patient monitoring via wearable devices, remote assistance via telemedicine, and remote analysis, all of which are important in providing crisis location, data the executives identified with drug, treatment, and clinical advice, and a cross authority mix of emergency clinic data frameworks.

Podpora *et al.* (2019) constructed a humanoid front desk office receptionist using an AI engine. The robot has the capability to engage visitors on various subjects relying on various sensors, and databases which allow it to become a useful interlocutor. The AI engine has learned to adapt to the bulk of real-world scenarios since the prototype was installed (Podpora *et al.* 2019).

Applications of IoT and robot in healthcare delivery are still evolving and it has been into several aspects of disease management and control. Some of the specific recent applications of IoT and robotic technology in the management of healthcare delivery are presented in Table 13.1. Figure 13.4 shows key features of an IoT-enabled robotic system. An IoT-driven robotic system is expected to have cognitive features and the ability to intelligently make decisions using connectivity.

TABLE 13.1
Recent Applications of IoT and Robotic Technology in Healthcare Delivery

Technology	Target Disease(s)	Application
Robotic technology	Medical receptionist	The robotic receptionist was created to engage with patients naturally at the beginning and end of a clinic appointment. A total of 40 individuals related with the robot during four exchanges to learn about their views of it and the results of 40 individuals evaluated the robot. The results of the human–robot interaction (HRI) after several interactions with the robot, shows that it may be a welcoming receptionist (Sutherland *et al.* 2019).

(*Continued*)

TABLE 13.1 (CONTINUED)

Technology	Target Disease(s)	Application
IoT wearable technology	COVID-19	COVID-19-related vital signs are measured and analysed using an IoT-based wearable monitoring device (Al Bassam *et al.* 2021).
IoT wearable technology and sensor devices	Heart disease	The focus was on monitoring health systems continuously using IoT wearable devices. Due to continuous data gathering, a scalable three-tier architecture was adopted to handle the resulting large data (Kumar & Devi Gandhi 2018)
IoT wearable OMgaments	Cardiac arrhythmia disease	Wearable electrocardiogram (ECG) sensors in the shape of an OMgarment for monitoring of the long-term rhythm with increased sensitivity in detecting intermittent or subclinical arrhythmia. The OMgaments' signal quality and accuracy were comparable to the Holter monitoring device (Steinberg *et al.* 2019).
Wearable IoT aldehyde sensor, Bluetooth low energy (BLE), cloud-based informatics system	Paediatric asthma, bronchial asthma	The wearable sensor is made in the shape of a wristwatch and measures formaldehyde levels in the air from 30 ppb to 10 ppm using fuel cell technology and it can run continuously for a week without a recharge. The sensor sends data to an mobile device over Bluetooth low energy wirelessly (BLE) for analysis and necessary action (Li *et al.* 2019). The latest advances in asthma management involving remote monitoring technologies have indeed shown significant promise to improve the quality of patient care and outcomes by employing different sensors for monitoring vital health metrics, such as oxygen saturation, heart rate, and environmental conditions, transmitting data in real time to healthcare providers. Other systems, such as ASTMATEST and Keva365, enable the monitoring of different parameters, which include peak flow measurements and symptom management, both of which are associated with improved patient outcomes in relation to the disease process (Mehta et al., 2024). Additionally, digital technologies support patient engagement, with 99.1% of individuals participating in remote patient monitoring programs exhibiting high adherence to their monitoring plans, an essential factor in the management of chronic diseases like asthma (Bizanti, 2024).c. A wearable IoT device was proposed that can predict asthma triggers by continuously sensing atmospheric parameters capable of triggering asthma and alerting the patient of the need to inhale a specific dosage as well as advising the patient to change location to prevent complications (Gundu 2020).
Wearable IoT device, mental health sensor, machine learning, robotic technology	Parkinson's disease	Wearable IoT device monitoring mental health was used to collect patient brain activities and cell status to forecast brain functionality changes (AlZubi *et al.* 2020). Detection of Parkinson's disease-related gait problems was examined using 16 IoT-based wearable sensors attached to the feet of the subjects. The support vector machine (SVM) is used to distinguish between Parkinson's patients and healthy people based on the data acquired from the sensors. The study concluded that a single sensor on the right foot of the subject might be sufficient to differentiate Parkinson's disease sufferers from healthy subjects (Channa *et al.* 2019).

(*Continued*)

TABLE 13.1 (CONTINUED)

Technology	Target Disease(s)	Application
		Authors investigated the clinical outcomes of patients who had "asleep" deep brain stimulation (DBS) for Parkinson's disease treatment employing robot-assisted electrode administration In the management of Parkinson's disease, their results for robot-assisted DBS of the subthalamic nucleus show comparable results to conventional methods (Moran *et al.* 2021). Postural instability in people with moderate Parkinson's disease and the effectiveness of robotic postural control was examined. The researchers concluded that robot-assisted balance training could be a potential method for improving patients' postural stability (Spina *et al.* 2021).
		The study focused on how robot-assisted gait training (RAGT) affects gait automaticity, gait speed, and stability in Parkinson's disease patients. RAGT utilizing an exoskeleton-type robot did not enhance gait automaticity in subjects suffering from Parkinson's disease, despite improvements in walking pace and stability (Yun *et al.* 2021).
Wearable IoT device, machine learning	Autism	Authors suggested utilizing wearable IoT sensors that send data to data acquisition and classification servers via a Bluetooth interface, and machine learning predictive accuracy of 91% was reported (Siddiqui *et al.* 2021).

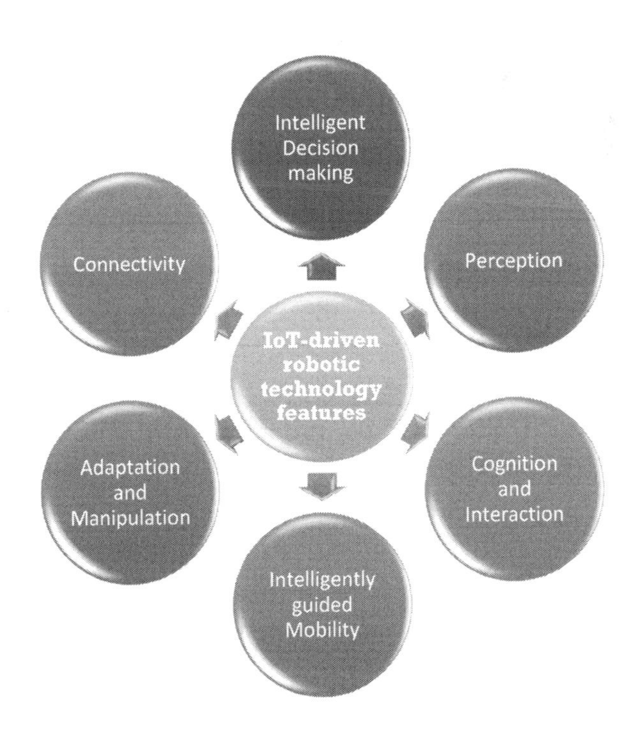

FIGURE 13.4 Key features of IoT-driven robotic technology.

CONCLUSION

The relevance of IoT, wearable technology, and robotics in medical healthcare delivery is presented with highlighted application in robotic receptionist and management and treatment of various diseases such as COVID-19, heart-related diseases like cardiac arrhythmia, paediatric asthma, bronchial asthma, Parkinson's disease and autism. The study showed that the future of medical healthcare delivery is closely connected to the use of various technologies that will improve how healthcare services and delivery are handled. IoT, wearables, immersible, sensors, and robotics are examples of the technologies that will redefine the future of medical health provision as the world gravitates towards realizing a smart earth. In the utilization of IoT and robotics technology, the healthcare sector has benefited greatly because it helps to lower the cost of healthcare delivery while also lowering the number of times of physical visits to doctors and emergency rooms.

REFERENCES

Adi, P.D.P. and Kitagawa, A., 2019. ZigBee Radio Frequency (RF) performance on Raspberry Pi 3 for Internet of Things (IoT) based blood pressure sensors monitoring. *International Journal of Advanced Computer Science and Applications*, 10 (5), 18–27.

Al Bassam, N., Hussain, S.A., Al Qaraghuli, A., Khan, J., Sumesh, E.P., and Lavanya, V., 2021. IoT based wearable device to monitor the signs of quarantined remote patients of COVID-19. *Informatics in Medicine Unlocked*, 24, 100588. 10.1016/j.imu.2021.100588

AlZubi, A.A., Alarifi, A., and Al-Maitah, M., 2020. Deep brain simulation wearable IoT sensor device based Parkinson brain disorder detection using heuristic tubu optimized sequence modular neural network. *Measurement: Journal of the International Measurement Confederation*, 161 107887.

Badenhop, C.W., Graham, S.R., Ramsey, B.W., Mullins, B.E., and Mailloux, L.O., 2017. The Z-Wave routing protocol and its security implications. *Computers & Security*, 68 112-129.

Beasley, R.A., 2012. Medical robots: Current systems and research directions. *Journal of Robotics*, 2012, 1–14.

Bizanti, S., 2024. Beyond the clinic: RPM's role in redefining asthma care. *Journal of Allergy and Clinical Immunology*, 153(2), AB361.

Channa, A., Baqai, A., and Ceylan, R., 2019. Design and application of a smart diagnostic system for Parkinson's patients using machine learning. *International Journal of Advanced Computer Science and Applications*, 10 (6), 563–571.

Chen, A.I., Balter, M.L., and Maguire, T.J., 2016. Developing the world's First portable medical robot for autonomous venipuncture [Industrial Activities]. *IEEE Robotics and Automation Magazine*. 23 (1), 10–11.

El Ouadghiri, M., Aghoutane, B., and El Farissi, N., 2020. Communication model in the internet of things. *Procedia Computer Science*, 177, 72–77.

Fan, X., Susan, F., Long, W., Li, S., Specification, Z., Gislason, D., Hillman, M., Fan, X., Susan, F., Long, W., Li, S., and Fan, B., 2017. Security analysis of zigbee. *MWR InfoSecurity*, (May) https://courses.csail.mit.edu/6.857/2017.project/17.pdf

Ghori, M.R., Wan, T.C., and Sodhy, G.C., 2020. Bluetooth low energy mesh networks: Survey of communication and security protocols. *Sensors (Switzerland)*, *20*(12), 3590.

Ginoya, T., Maddahi, Y., and Zareinia, K., 2021. A historical review of medical robotic platforms. *Journal of Robotics*, 2021, 6640031

Gomez, C., Oller, J., and Paradells, J., 2012. Overview and evaluation of bluetooth low energy: An emerging low-power wireless technology. *Sensors (Switzerland)*, 12 (9), 11734–11753.

Gundu, S., 2020. A Novel IoT based solution for monitoring and alerting bronchial asthma patients. *International Journal of Research in Engineering, Science and Management*, 3 (10), 120–123.

Hussein, Z.K., Hadi, H.J., Abdul-Mutaleb, M.R., and Mezaal, Y.S., 2020. Low cost smart weather station using Arduino and ZigBee. *Telkomnika (Telecommunication Computing Electronics and Control)*, 18 (1), 282–288.

Johanson, D.L., Ahn, H.S., Sutherland, C.J., Brown, B., MacDonald, B.A., Lim, J.Y., Ahn, B.K., and Broadbent, E., 2020. Smiling and use of first-name by a healthcare receptionist robot: Effects on user perceptions, attitudes, and behaviours. *Paladyn*, 11 (1), 40–51.

Kumar, P.M. and Devi Gandhi, U., 2018. A novel three-tier Internet of Things architecture with machine learning algorithm for early detection of heart diseases. *Computers and Electrical Engineering*, 65, 222–235.

Li, B., Dong, Q., Downen, R.S., Tran, N., Jackson, J.H., Pillai, D., Zaghloul, M., and Li, Z., 2019. A wearable IoT aldehyde sensor for pediatric asthma research and management. *Sensors and Actuators, B: Chemical*, 287, 584–594.

Makarchuk, R., 2022. Is LoRaWAN going to drive 5G out of the IoT business? - Intellias [online]. Available from: https://intellias.com/the-telecom-dilemma-making-the-most-of-lorawan-and-5g-to-power-the-iot-landscape/ [Accessed 9 Mar 2022].

Mehta, J., Garrett, J., Patil, R., and Badran, J., 2024. Patient engagement on a remote therapeutic monitoring program for chronic disease management of asthma. *Journal of Allergy and Clinical Immunology*, 153(2), AB182.

Mettler, L., Ibrahim, M., and Jonat, W., 1998. One year of experience working with the aid of a robotic assistant (the voice-controlled optic holder AESOP*) in gynaecological endoscopic surgery. *Human Reproduction*, 13 (10), 2748–2750.

Moran, C.H., Pietrzyk, M., Sarangmat, N., Gerard, C.S., Barua, N., Ashida, R., Whone, A., Szewczyk-Krolikowski, K., Mooney, L., and Gill, S.S., 2021. Clinical outcome of "asleep" deep brain stimulation for Parkinson disease using robot-assisted delivery and anatomic targeting of the subthalamic nucleus: A series of 152 patients. *Neurosurgery*, 88 (1), 165–173.

Nguyen, P.T., Lorate Shiny, M., Shankar, K., Hashim, W., and Maseleno, A., 2019. Robotic surgery. *International Journal of Engineering and Advanced Technology*, 8 (6 Special Issue 2), 995–998.

O'Hara, J., 2007. Toward a commodity enterprise middleware: Can AMQP enable a new era in messaging middleware? A look inside standards-based messaging with AMQP. *Queue*, 5(4), 48–55.

OM Signal Inc., 2015. OMsignal biometric smartwear. *OMsignal*.

Oniani, S., Marques, G., Barnovi, S., Pires, I.M., and Bhoi, A.K., 2021. Artificial intelligence for internet of things and enhanced medical systems. *Studies in Computational Intelligence*, 43–59.

Park, J., Bhat, G., Anish, N.K., Geyik, C.S., Ogras, U.Y., and Lee, H.G., 2020. Energy per operation optimization for energy-harvesting wearable IoT devices. *Sensors (Switzerland)*, 20 (3) 764.

Pereira, D.S., De Morais, M.R., Nascimento, L.B.P., Alsina, P.J., Santos, V.G., Fernandes, D.H.S., and Silva, M.R., 2020. Zigbee protocol-based communication network for multi-unmanned aerial vehicle networks. *IEEE Access*, 8, 57762–57771.

Petrescu, V.V.R., 2019. Medical service of robots. *Journal of Mechatronics and Robotics*, 3 (1), 60–81.

Podpora, M., Gardecki, A., and Kawala-Sterniuk, A., 2019. Humanoid receptionist connected to IoT subsystems and smart infrastructure is smarter than expected. *IFAC-PapersOnLine*, 52(27), 347–352.

Pušnik, M., Galun, M., and Šumak, B., 2020. Improved Bluetooth low energy sensor detection for indoor localization services. *Sensors (Switzerland)*, 20 (8) 2336.

Shelby, Z., Hartke, K., and Bormann, C., 2014. The constrained application protocol (CoAP). *Internet Engineering Task Force* (IETF) RFC-7252. http://dx.doi.org/10.17487/RFC7252

Shiraishi, S., 2013. 3. WebSocket. *The Journal of the Institute of Image Information and Television Engineers*, 67 (2), 102–108.

Siddiqui, U.A., Ullah, F., Iqbal, A., Khan, A., Ullah, R., Paracha, S., Shahzad, H., and Kwak, K.S., 2021. A wearable-sensors-based platform for gesture recognition of autism spectrum disorder children using machine learning algorithms. *Sensors*, 21 (10) 3319.

Sinha, V., Malik, M., Nugent, N., Drake, P., and Cavale, N., 2021. The role of virtual consultations in plastic surgery during COVID-19 Lockdown. *Aesthetic Plastic Surgery*, 45 (2), 777–783.

Spina, S., Facciorusso, S., Cinone, N., Armiento, R., Picelli, A., Avvantaggiato, C., Ciritella, C., Fiore, P., and Santamato, A., 2021. Effectiveness of robotic balance training on postural instability in patients with mild Parkinson s disease: A pilot, single-blind, randomized controlled trial. *Journal of Rehabilitation Medicine*, 53 (2), 1–9.

Stanford, 2022. Robotics: A brief history [online]. Available from: https://cs.stanford.edu/people/eroberts/courses/soco/projects/1998-99/robotics/history.html [Accessed 13 Feb 2022].

Steinberg, C., Philippon, F., Sanchez, M., Fortier-Poisson, P., O'Hara, G., Molin, F., Sarrazin, J.F., Nault, I., Blier, L., Roy, K., Plourde, B., and Champagne, J., 2019. A novel wearable device for continuous ambulatory ECG recording: Proof of concept and assessment of signal quality. *Biosensors*, 9 (1), 17.

Sutherland, C.J., Ahn, B.K., Brown, B., Lim, J., Johanson, D.L., Broadbent, E., Macdonald, B.A., and Ahn, H.S., 2019. The doctor will see you now: Could a robot be a medical receptionist? In: *Proceedings - IEEE International Conference on Robotics and Automation*. Institute of Electrical and Electronics Engineers Inc., 4310–4316.

Tavakoli, M., Carriere, J., and Torabi, A., 2020. Robotics, smart wearable technologies, and autonomous intelligent systems for healthcare during the COVID-19 pandemic: An analysis of the state of the art and future vision. *Advanced Intelligent Systems*, 2, 2000071.

Van den Eynde, J., De Groote, S., Van Lerberghe, R., Van den Eynde, R., and Oosterlinck, W., 2020. Cardiothoracic robotic assisted surgery in times of COVID-19. *Journal of Robotic Surgery*, 14 (5), 795–797.

Vinoski, S., 2006. Advanced message queuing protocol. *IEEE Internet Computing*, 10 (6), 87–89.

Wang, J., Peng, C., Zhao, Y., Ye, R., Hong, J., Huang, H., and Chen, L., 2021. Application of a robotic tele-echography system for COVID-19 pneumonia. *Journal of Ultrasound in Medicine*, 40 (2), 385–390.

WebSockets Standard [online], 2022. Available from: https://websockets.spec.whatwg.org// [Accessed 14 Mar 2022].

WWS - Garment [online], 2022. Available from: https://www.smartex.it/en/our-products/wws-components/251-wws-garment [Accessed 9 Mar 2022].

Xiao, J. and Li, J.T., 2020. Design and implementation of intelligent temperature and humidity monitoring system based on ZigBee and WiFi. *Procedia Computer Science*, 166, 419–422.

Yildirim, H. and Ali-Eldin, A.M.T., 2019. A model for predicting user intention to use wearable IoT devices at the workplace. *Journal of King Saud University - Computer and Information Sciences*, 31 (4), 497–505.

Yun, S.J., Lee, H.H., Lee, W.H., Lee, S.H., Oh, B.-M., and Seo, H.G., 2021. Effect of robot-assisted gait training on gait automaticity in Parkinson's disease. *Medicine*, 100 (5), e24348.

Zeng, Z., Chen, P.J., and Lew, A.A., 2020. From high-touch to high-tech: COVID-19 drives robotics adoption. *Tourism Geographies*, 22 (3), 724–734.

Zhang, M., Wang, X., Feng, H., Huang, Q., Xiao, X., and Zhang, X., 2021. Wearable Internet of Things enabled precision livestock farming in smart farms: A review of technical solutions for precise perception, biocompatibility, and sustainability monitoring. *Journal of Cleaner Production*, 312, 127712.

Zradziński, P., Karpowicz, J., Gryz, K., Morzyński, L., Młyński, R., Swidziński, A., Godziszewski, K., and Ramos, V., 2020. Modelling the influence of electromagnetic field on the user of a wearable IoT device used in a WSN for monitoring and reducing hazards in the work environment. *Sensors (Switzerland)*, 20 (24).

Z-Wave Alliance, 2021. About Z-wave technology - Z-wave alliance [online]. *Z-Wave Alliance*. https://z-wavealliance.org/about_z-wave_technology/

14 The Role of 5G in Healthcare Centres Towards the Provision of Healthcare

Akinola Samson Olayinka
Edo State University, Uzairue, Nigeria

Onyijen Ojei Harrison
Samuel Adegboyega University, Ogwa, Nigeria

Tosin Comfort Olayinka
College of Computing, Wellspring University,
Benin City, Nigeria

Charles Oluwaseun Adetunji
Edo State University, Uzairue, Nigeria

Clement Atachegbe Onate
Landmark University, Omu-Aran, Nigeria

Olusanmi Ebenezer Odeyemi
Federal College of Animal Health and
Production Technology, Ibadan, Nigeria

Oluwafemi Adebayo Oyewole
Federal University of Technology, Minna, Nigeria

K. I. T. Eniola
Joseph Ayo Babalola University, Ikeji Arakeji, Nigeria

 DOI: 10.1201/9781003309468-14

14.1 INTRODUCTION

Technological innovations and development in the areas of wireless communication were part of the technological breakthroughs of the last few years, especially with the introduction of the much-debated fifth-generation (5G) wireless spectrum (Ndinojuo 2020; Cheng et al. 2021). The research interest in wireless communication sector has always been tailored towards achieving a better speed at a reduced latency (Belikaidis et al. 2017, Olayinka et al. 2018, n.d.; Sachs et al. 2019; Kiesel et al. 2020; Nwankwo et al. 2020). Remote healthcare delivery can only be achieved where the high speed with low latency communication can be guaranteed as in 5G. 5G wireless communication is an improvement on the fourth-generation (4G) long-term evolution (LTE) wireless communication. Wireless communication has been used in various aspects of healthcare delivery such as telemedicine, telesurgery, telecare, teledentistry, telenursing, telehealth, teleconsultation, and telepharmacy among others and it became more popular and relevant with the COVID-19 pandemic (Gil et al. 2007; Calton et al. 2020; Doraiswamy et al. 2020; Gogia 2020; Melton et al. 2020; Paterson et al. 2020; Wong et al. 2020). Another important area of wireless communication is inter-departmental data exchange locally and remotely within and outside the healthcare facilities, e.g. laboratory test results transmitted wirelessly to doctors, and doctors' recommendations and prescriptions are transmitted to nursing and pharmacy departments. One of the major challenges facing most rural communities in developing nations of the world is in the aspect of healthcare delivery. This is largely due to the inadequate number of medical personnel to attend to the large population in such locations.

5G wireless communication will play a major role in revolutionizing the way things are done in the years to come. With rapid development in quantum computing and 5G network capability, many developments and breakthroughs are envisaged in the internet of things (IoT), internet vision, smart vehicles, smart homes, smart medicine, smart healthcare, smart agriculture.

A healthcare system where technology will play a dominant role in the delivery of services to users is often referred to as smart healthcare. Relevant technology in smart healthcare includes IoT, biosensors, wearable technology, as well as high-speed wireless communication such as 5G for dynamic synchronous access to information. 5G communication also ensures a connected system and users within the domain of the healthcare system, while proving a suitable medical ecosystem that can respond and intelligently manage healthcare service (Sachs et al. 2019, Tian et al. 2019). Malaria, for instance, is still a major health challenge in sub-Saharan Africa with children being most affected; however, 5G, IoT, machine learning, biosensor technology as well as improved nanoparticles-based drug design can help reduce the casualty rate (Olayinka and Chiemeke 2019; Dutta 2020; WHO 2020; Adetunji et al. 2021; Prevention 2021). In this work, progress and developments in 5G wireless communication's potential in improving quality healthcare delivery are presented.

14.2 WIRELESS NETWORK EVOLUTION FROM 1G TO 6G

It has been over four decades since the first generation (1G) of wireless technology was unveiled. Over the years, mobile phones have become smaller, internet speed has improved, social media apps have been used more than ever and time spent on the internet by individuals has increased globally. The timeline of the telecommunication generation that began in 1979 has revolutionized from first generation (1G) to fifth generation (5G).

The 1G mobile communication system was first launched in 1979 by the Nippon Telegraph and Telephone (NTT) Corporation in Japan and by 1984, the whole of Japan had access to 1G mobile networks (Kikuchi et al. 1985). The 1G mobile network was approved and rolled out in the US by Motorola in 1983 followed by Canada and the UK years later. The major setback with the 1G mobile network was the poor quality of sound, inadequate coverage, and insecurity. Despite the setbacks, Motorola's DynaTAC, the first commercial mobile phone was able to achieve 20 million subscribers globally by the end of 1990, which was a motivation for the development of the second generation (O'Regan 2016, Galazzo 2022, Haverans 2022, Saravia 2022).

The second-generation (2G) network was introduced sometime in 1990 and it was a paradigm shift from the analogue, 1G to a digital terrain of 2G. It was altogether an improvement to the existing 1G in terms of audio quality by binary digits (bits) error correction of the audio signal. The International Telecommunications Union (ITU) had an active role in organizing the telecommunication sector across different parts of the world. The 2G network also heralded global access to personal mobile internet with spectrum progressions from the successes in 1G. The following developments were made with the communication spectrum; GPRS (General Packet Radio Service), GSM (Global Systems for Mobile Communication), and EDGE (Enhanced Data rates for GSM Evolution) with approximate data speeds of 30–35 kbps, 110 kbps and 135 kbps respectively. EDGE was unveiled in 2003, and it was a significant improvement over GSM and GPRS. EDGE is sometimes referred to as 2.9G. It is still used by many mobile network providers to date.

The third generation (3G) mobile wireless network was a major milestone in the history of mobile data communication with a speed potential of up to 2 mbps. Faster mobile communication was enabled for the exchange of large email files and text messages, streaming of video online, high-speed internet browsing and improved security protocol. The driving technology is EDGE and CDMA (Code-Division Multiple Access).

The fourth generation (4G) came with several standards which include data exchange over internet protocol (IP), IP telephony and voice over IP, video conferencing, high-definition (HD) mobile TV, and a data speed of at least of 100 Mbps. 4G is sometimes defined with the acronym MAGIC with M meaning "Mobile multimedia"; A meaning "Anytime anywhere"; G meaning "Global mobility support"; I meaning "Integrated wireless solution" and C meaning "Customized personal service" (Anitha et al. 2015; Sabah 2016).

Fifth-generation (5G) mobile network communication is still evolving globally, Though South Korea claimed to have been the first country in the world to launch a 3.5 GHz band 5G network in 2019 (Lee et al. 2021), the 5G network is still not available in several countries of the world. Efforts are being geared towards improving the quality of service globally with South Korea adopting the slogan "first in the world in the race to 5G to first in global quality" (Gillispie

2020). Researchers in South Korea have smoothed the tension over radiation safety on the 5G network. Electromagnetic fields exposure was measured for communication antennas (2G, 3G, 4G and 5G) in South Korea and the results showed only about 15% of the emission is traceable to the 5G network. The study concluded that peak radiation measured was below the international commission standard on non-ionizing radiation (Selmaoui et al. 2021). The 5G network comes with a faster speed of up to 10 Gbps and minimal power requirements, which makes it suitable for application in the internet of things (IoT), big data, and internet of medical things (IoMT).

5G introduces new capabilities and opportunities such as massive multiple-input, multiple-output (MMIMO) networks, network slicing, and orthogonal frequency division multiplexing (OFDM). 5G will also usher in a new standard known as 5G New Radio (NR), which will eventually replace LTE. 5G NR expands on LTE's finest features and adds new features such as B. Increased power savings and increased communication for connected devices.

Furthermore, 5G can use a new frequency spectrum called millimeter wave (MM wave), which uses wavelengths between 30 and 300 GHz, whereas 4G LTE uses wavelengths below 6 GHz. New tiny cell base stations are necessary for 5G operation and function due to the MM wave spectrum.

Sixth-generation (6G) wireless technology is still a subject of research by various experts in the telecommunication industry. It is expected to be unveiled on or before the year 2030. When unveiled, it is expected to provide a data speed of up to 1 Tbps with latency less than 1 ms. The 6G network will fully support the developments in artificial intelligence, autonomous vehicle, haptic communication, terahertz (THz) communication and full satellite integration. Some of these developments are partly supported by the 4G and 5G networks. The 5G and 6G wireless standards are networks of the future because they are still the subject of active research (Figure 14.1).

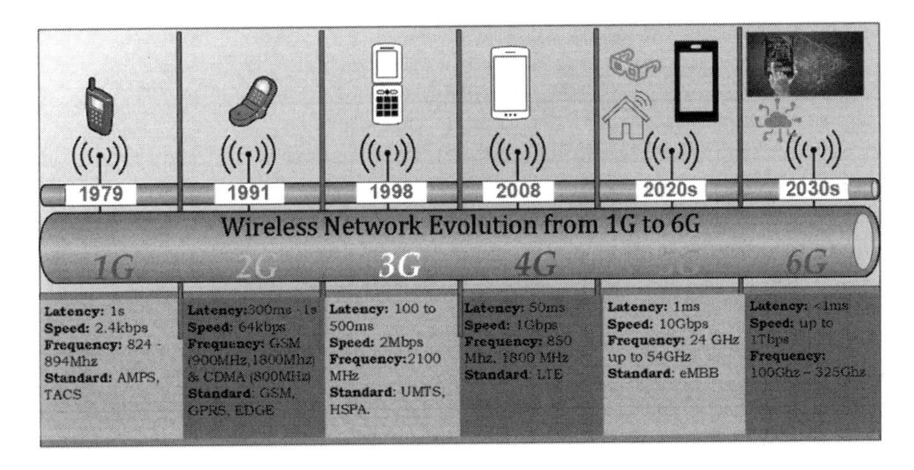

FIGURE 14.1 Evolution of mobile communication from 1G to 6G.

14.3 APPLICATION OF 5G IN HEALTHCARE DELIVERY

Healthcare delivery is exponentially growing very fast and expanding, with an enormous increase in the number of applications using the Internet. Much data are generated from the healthcare industry and they are made up of numerous data types, formats, and sizes. As a result of this, it places compound demands over the network related to the data rate, latency, bandwidth, together with other related factors. Over time, the e-healthcare service is deployed at different healthcare centres with the use of network technologies for the connection of possible networks. The network technologies include Wi-Fi, Bluetooth, and ZigBee among others. With the improvement of e-healthcare delivery, there is a need to improve the connecting technologies, since there is an higher number of devices and sensors to activate the sensor applications concurrently in a big e-health.

Recently, 5G technologies are integrating the world into a singular system such that it has made important innovations in several real-world applications. 5G-based technology has been introduced in energy conservation, healthcare, and also in commercial fields. 5G has created several new opportunities in healthcare in the areas like Massive-Machine Type Communication (mMTC), data analysis, medical imaging, and prognosis because of its extremely low latency, intelligent health administration, as well as how medical Big Databases are handled (West 2009; Newswire 2016). 5G technologies can enhance network capacity due to their high speed, low latency, increased throughput, and reduced packet loss (Mishra et al. 2021).

5G is the cellular wireless technology that has come to existence through fifth-generation advancements, which have the potential to have the strength of a large network combined with very high speed. This can enable timely delivery of quality e-healthcare services because an interruption of a fraction of picoseconds can aggravate the situation. 5G technology is relevant and important for the implementation of the IoMT. IoMT includes virtual reality, remote patient monitoring, artificial intelligence, remote healthcare learning, augmented reality, and many other related technologies (Ghazal 2021; Tarikere et al. 2021). 5G also enables essential technology in the medical IoT field, in order to access real-time data and make instant decisions from the data captured. Some characteristics of 5G technology communication include; communication speed, response time, connection density, survival time, user mobility support, universal application, support, security, and privacy. These features are useful in developing an efficient e-healthcare system and maintaining the usability and storage of the patients' data in a protected way.

In Africa, healthcare systems have been overwhelmed with inefficiencies, and this leads to ill health and a cycle of poverty. However, there has been improvement in the clinical management of various diseases like malaria, diabetes, hypertension, and tuberculosis. Digital health involves clinical communication, virtual visits, and e-health, which represents services that are delivered through electronic means like the internet and mobile devices in case of m-health (Laukka et al. 2021).

Ashleibta et al. (2021) studied a 5G-enabled contactless multi-user assisted living healthcare system. In their study, they presented the foremost 5G-enabled contactless RF-sensing system which operates at about 3.75 GHz for multi-subject, useful for monitoring in-home health activity. The system will monitor the presence of humans and also sense parallel multi-user activity that relies on channel state information (CSI) signals which focuses on a small setting, like rooms in a care home, with four persons. The system basically performs Sitting, Standing and Walking activities. A combination of a convolutional neural network and deep learning approach was used to test and trained on the channel state information data to classify activities of the subject. The result shows that the proposed system has an average accuracy of 91.25%, which is high for sole object movements, and also has the highest multi-class accuracy of 83% for 4 subjects and 16 classification categories. The system has the potential to monitor in-home health activity and for the purpose of keeping track of the public's health.

Tian et al. (2019) reviewed smart healthcare and how to make medical healthcare more intelligent. In their review, they stated the generated novel information technology is useful for smart healthcare. The novel information technologies like artificial intelligence, the internet of things (loT), cloud computing, and big data have transformed all facets of the traditional healthcare system. Health management, supporting diagnosis and treatment, illness prevention and risk control, virtual assistants, smart hospitals, and assisting drug development are all areas where smart healthcare can be used (Rath and Pattanayak 2019; Ahad et al. 2020). The benefits of smart healthcare include reducing costs, improving the patient's medical experience, relieving personnel pressure and creating a unified material and information management. Despite the emergence of new technologies, so many challenges are now emerging. The challenges include the inadequate provision of macro programmatic documents, the non-existence of uniform standards among different medical institutions, and insufficient data integrity.

Li et al., 2020 presented a study on the perception of various stakeholders on the adoption of e-hospitals in Western China. The study provided insights about the approval of e-hospitals by examining patients' willingness on the usage of e-hospitals and also analyzed the barriers to embracing e-hospital technology. Questionnaires were administered for 1032 patients and analyzed cross-sectionally using multivariate logistic regression. Patients' sociodemographic characteristics collected were medical history, skillful usage of electronic devices, web-based health services experience, current disease status, readiness to use e-hospitals, and lastly, apparent barriers and facilitators. The result suggested that there should be increased efforts for the adoption of e-hospitals.

Shu and Li, (2020) presented China's experience with digital health in the COVID-19 pandemic. In their study, they reported four approaches those Chinese hospitals have implemented to tackle the problem. These approaches are telemedicine services, intelligent robots, contact-free strategy, and community outreach and support. Telemedicine services have provided free online medical consultation services immediately after the outbreak. Telemedicine encompasses the usage of telecommunication to offer healthcare information and related services (Olayinka et al. 2018). Telemedicine combines the knowledge of telecommunication concept and medical practices for remote healthcare services (Setyo et al. 2019). This system uses satellites to provide video conference consultations to healthcare facilities. The planned clinic for outpatients was transferred online or used on mobile apps, which have been promoted by online specialist consultation services. Several Chinese health workers were at a high infection risk. To relieve the health workers some Chinese hospitals installed intelligent robots. These intelligent robots are useful in the infectious disease sections of hospitals, to reduce the danger of transferring infection and preserve valuable resources. The robots are used for disinfecting facilities, distributing goods and also providing scrutiny in hospitals. The emergence of telemedicine systems and intelligent robotics has enabled the provision of contactless medical services (Global Times, 2020; Beijing News, 2020). To collect and transmit patients' information in some segregated wards, remote health-monitoring devices, cloud video services, and mobile telecom devices were used. Moreover, innovative contact-free digital technology was used for fever detection, drug refill process, and isolation wards care. There is community social collaborative support for the digital technology which enables specific hospitals to provide treatment for fever and coronavirus diseases (Lian et al. 2020).

Ahad et al. (2020) reviewed how technological advancement towards 5G will impact smart healthcare delivery using IoT. In the future, 4G communication technology may struggle to meet the needs of highly dynamic and time-sensitive healthcare applications. The 5G network was developed to provide possible solutions to various communication healthcare applications needs with the IoT. This technology is very useful in smart healthcare system networks. The paper presented a detailed assessment of 5G-aided smart healthcare solutions with IoT, smart healthcare structure for 5G, the essential requirements for effective smart healthcare systems deployment for some cases and lastly some shortcomings encountered in 5G smart healthcare solutions.

Ullah et al. presented an architectural framework for body area networks (BANs) that can support smart delivery of healthcare services by utilizing a wakeup-radio protocol for data collections (Ullah et al. 2019).

Hu et al. developed a prototype 5G-enabled secure smart healthcare delivery system to monitor hypertensive related heart disease. Experimenting with the prototype with 45 subjects with cardiovascular disease (CVD), the system gave 96.34% accuracy with a sensitivity and specificity of 92.46% and 93.62% respectively (Hu et al. 2021).

Hameed et al. used internet of health things (IoHT) powered medical devices for data collection from patients to perform a smart diagnosis. 5G wireless network and blockchain technology were deployed for cloud data storage while a neural network classifier was used for disease prediction (Hameed et al. 2021).

Mishra et al. (2021) studied the use of IoMT powered by 5G NR network for effective health monitoring. The paper presented the principal and crucial features of 5G technology communication, and the taxonomy of the 5G network which can be applied to provide real-time solution healthcare domain. The 5G communication technologies can be short-range and long-range with the 5G performance metrics. 5G technology has a ten times lower latency than 4G technology and can accommodate a larger number of devices. With 5G NR, there is access to an improved electronic healthcare facility by sharing data and detecting disease which becomes faster and easier. 5G is applicable in remote patient monitoring (Griggs et al. 2018; Malasinghe et al. 2019; Annis et al. 2020; Gordon et al. 2020), connected ambulance (Elsaadany et al. 2018; Alami-Kamouri et al. 2019; Usman et al. 2019), virtual consultation (Kilvert et al. 2020; Gilbert et al. 2021; Murthy et al. 2021; Proulx-Cabana et al. 2021; Sinha et al. 2021), real-time maintenance (Huang et al. 2018; Mourtzis et al. 2020), augmented reality (AR) technology and virtual reality (VR) technology (Desselle et al. 2020; Muñoz-Saavedra et al. 2020; Singh et al. 2020), robotic surgery (Nguyen et al. 2019; Kamruzzaman 2020), cloud computing (Kumari et al. 2020). 5G technologies can also offer a wider range of services to commercial enterprises, clients, and third-party service providers (Brown 2018) (Table 14.1).

TABLE 14.1
Application of 5G in Healthcare Delivery

s/n	Technology	Application	References
1	5G, big data, clouds	Smart continuous monitoring of diabetic disease.	Chen et al. (2018)
2	5G, big data	Smart medical and nursing platform for medical information exchange among medical personnel, vital signs smart monitoring, pension management support as well as decision support system-making system and telemedicine diagnosis.	Liu et al. (2021)
3	5G, AI, IoT, cloud computing and virtual reality.	The use of an intelligent medical system, 5G enabled smart medicine and smart health management to prevent sickness by encouraging a healthy lifestyle.	Ying and Yu (2021)
4	Unmanned aerial vehicle (UAVs), wakeup-radio protocol, body area network (BANs), internet of medical things (IoMT)	UAV-supported architecture to collect data from BANs. This is architecture is believed to be relevant in distant or disaster-stricken locations where BANs have limited or no access to traditional wireless communication infrastructure.	Ullah et al. (2019)
5	5G-enabled contactless RF-sensing system, deep learning	5G-enabled system for in-home smart health monitoring with potential for human activity recognition (HAR) for assisted living.	Ashleibta et al., (2021)
6	5G, wireless network	Internet of medical things (IoMTs) for smart health monitoring.	Mishra et al., (2021)
7	AI, 5G network, big data	5G-powered smart hospital for improving healthcare delivery through hospital administration and procedures such as diagnosis, patient record, online payments, etc.	Li et al., (2021)
8	5G technology for e-health	Application of mobile devices, remote monitoring devices and mobile IoT (mIoT) in telemedicine as well as management and tracking of hospital facilities like wheelchairs, electrocardiogram (ECG) monitors, X-ray systems.	Padmashree and Nayak, (2020)
9	5G, augmented reality (AR), UAVs, robotic technology and wearable technology	Telehealth application in telemedicine, telenursing, telesurgery and telepharmacy using 5G-enabled system as well as management of COVID-19 pandemic.	Siriwardhana et al. (2021)
10	Wireless network	Internet of things (IoT) base smart healthcare delivery.	Ahad et al., (2020)
11	Wireless network	Web-based and desktop-based healthcare access in telemedicine system	Olayinka et al., (2018)
12	5G, 5G-IPv6, blockchain	Smart health secure monitoring of hypertensive heart disease. Stages include data collection, diagnosis and health service.	Hu et al. (2021)
13	5G, blockchain, IoT, neural network	Using the internet of health things (IoHT) to power medical devices to collect data of patients. Data is stored in the cloud server for a neural network classifier to predict disease and determine the severity of the disease.	Hameed et al. (2021)

14.4 CHALLENGES OF 5G IN SMART HEALTHCARE

a. **Accomplishing Interoperability**
Interoperability involves the ability to connect various devices and networks for effective data exchange. Interoperability provides a platform whereby various devices are connected with different communication systems. The interoperability among several domains become a challenge for IoT success due to a lack of universal standards communications technologies (Noura et al. 2019). For this reason, there is a need for an intelligent method to check the interoperability at diverse levels and enable several devices within the network to effectively communicate.

b. **Big Data Analysis**
Analysis of big data is one of the foremost research routes in smart healthcare networks. With the smart healthcare network, several million devices will generate a massive amount of data and information that may be analyzed (Ahad et al. 2020). These data comprise private user information and the immediate environment patient's information. Consequently, intelligent approaches are needed significantly for data analysis.

c. **IoT Connectivity**
The smart healthcare network can be successful if every device is rightly connected. The devices provide information after sensing is carried out. Here, any existing communication network system can be used by the IoT devices, like a mobile network, Wi-Fi and Bluetooth. Nevertheless, there are several challenges to ensuring that each device in the smart healthcare network is well connected. Assuring connectivity to devices with high mobility in the network, as well as providing connectivity to every device placed in the network, are two of these problems.

d. **Accomplishing Security, Trust, and Privacy**
Security is a serious challenge in smart healthcare networks, which is due to the interconnection of various IoT devices. Deploy multifaceted security algorithms and protocols are difficult due to processing power on IoT devices and limited battery life. There is a tendency that in future, IoT devices will be at the peril of attacks, leading to different security threats and attacks as well as confidentiality issue. In designing an effective 5G network enabled smart healthcare system, the following should be considered:
   i. Data validity, integrity, and security as well as clear connection between smart healthcare equipment and cloud database centres must be guaranteed.
   ii. A definite approach for risk profiling and assessment must be provided, to proactively identify attacks.
   iii. Provision of the standard privacy policy for new users access and confidentiality must be guaranteed.

14.5 FUTURE ADVANCEMENT OF 5G IN HEALTHCARE

The application of 5G in the healthcare system has the potential to improve data communications and rapid transmission of medical big data for rapid diagnosis of various diseases and management of diseases. 5G promises lower latency than its predecessor, 4G and higher speed to accelerate disease diagnosis and therapy. Higher bandwidth, ultra-low latency, and ultra-reliability will all be part of the coming 5G era, as will the growing adoption of cloud-based storage and a variety of linked devices. Currently, some technological barriers (such as poor data rates, limited connectivity, security issues, and so on) are preventing the widespread acceptance of home-based healthcare monitoring systems. By delivering enhanced security, better bandwidth, reliable transmission, and uninterrupted accessibility, the future 5G networks based infrastructure will open up novel alternatives to address these limits. This will give health-monitoring services that are constantly available, with a significant increase in transmission capabilities between healthcare providers and patients (Latif et al. 2017). Figure 14.2 shows the key features that necessitate the need for 5G network as a prefer network in healthcare delivery. The applications of 5G network in healthcare delivery is shown in Figure 14.3. The applications areas include smart remote monitoring of diseases, medical connected devices, and medical information exchange. Other relevant application areas are augmented reality in delivery of healthcare and provision of medical services through telemedicine, telenursing, telesurgery and telepharmacy.

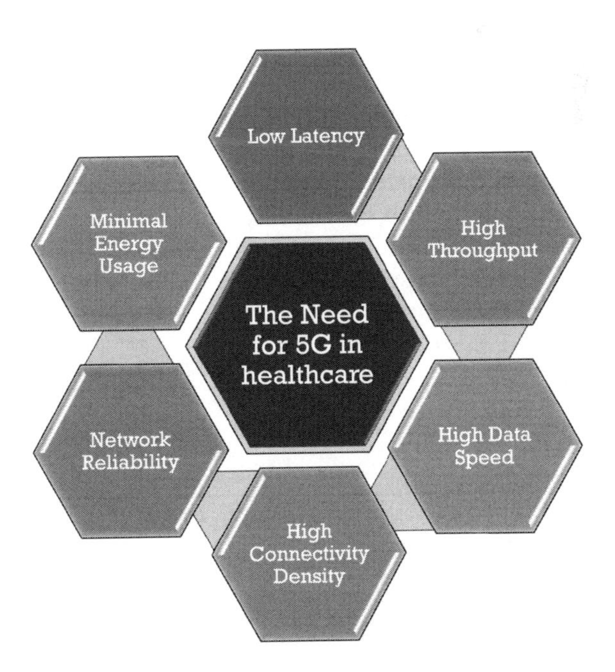

FIGURE 14.2 The need for 5G in healthcare delivery.

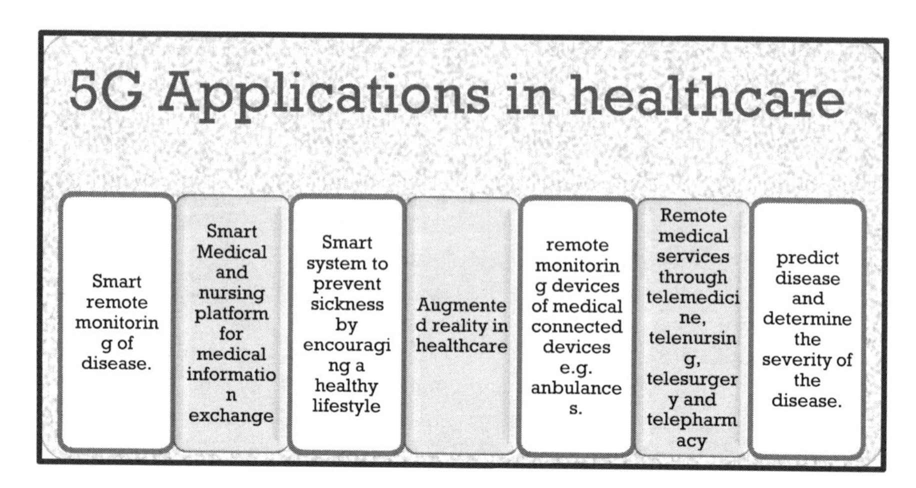

FIGURE 14.3 The applications of 5G in healthcare delivery.

14.5.1 Imaging

Digital medicine has two advantages that help to compress time and distance: remote image access and the capacity to quickly transmit data across geographic boundaries. If a physician from another part of the country or the world needs a second opinion, they can email a medical image or test result to another doctor for their input. This enables clinicians to gain requisite experience while also assisting other healthcare system to address imbalances based on geographical location, financial constraint, or socioeconomic class.

People in these situations usually don't have access to current medical information; however, they may benefit from the expertise of specialists who practise far away thanks to digital technology. As a result, health disparities are reduced, and the urban/rural split in most nations is bridged. Patients are not required to travel physically to receive high-quality medical care. Patients can get second or third views rapidly because of the faster transmission of the images from X-rays or CT scanning machines. The barrier of restriction to specialists in their locality is eliminated, by having access to several medical experts globally. This will benefit patients by expanding the skill pool and bringing highly responsive health expertise to tiny towns with few health facilities.

14.5.2 Diagnostics

As the usage of monitoring devices and wearable medical equipment expands, advancements in diagnostics are becoming increasingly vital. For patients with major health problems like diabetes, cancer, and heart disease among others, remote healthcare monitoring tools can be very useful to monitor parameters like the vital signs, the blood pressure, and the glucose levels, and then transmit this information electronically to healthcare experts for relevant actions. Rather than waiting for an emergency, these devices and immersive connectivity provide an early warning system that allows doctors to notice possible problems and treat patients before they become serious. This feature is already available on existing 4G communication networks and equipment, but what sets 5G apart is its ability to deliver machine-to-machine communications, which will help with real-time monitoring, control and data analytics.

The Foundation by the M. J. Fox was a forerunner in the development of the equipment that tracks the tremors associated with Parkinson's disease. Wearable sensor devices are being used by medical experts to acquire reliable real-time data for different aspects of the disease, instead of relying on patients' personal report of tremor frequency and duration, as well as how they have changed over time. This volume of data is unparalleled, and the capacity to analyze it and spot trends will aid in evaluating whether symptoms are worsening and what might be causing them. People will be able to seek medical aid through video conferencing and telemedicine as 5G advances in healthcare. Medical personnel can monitor several parameters to diagnose people's health concerns in real-time (Hume and Jeff 2022).

There have already been significant advancements in the ability to undertake in-home health monitoring practically. Individuals can monitor their vital sign using wearable devices at a lower cost and with greater convenience than in the past, which required a trip to the local hospital. Developments in 5G wireless technology has the potential to motivate a devolution in the way healthcare services are accessed in homes, surgical theatre, medical facilities and remote locations as well as mobile facilities like ambulances. 5G will also provide new potential for radar technology for non-intrusive monitoring in both the home and hospital settings (Latif et al. 2017).

In Taiwan, for example, the city of Taipei has implemented the Citizen Telecare Service System (CTSS), a system for handling healthcare data. The government is attempting to overcome geographical limits, redistribute medical facilities, and offer advanced individuals with a sense of comfort while keeping an eye on their bodily state by using a telecare information platform. The program's goal is to fully integrate technology that enables continuous monitoring of biometric data, tracking, and early warnings associated with deviant health situations. Education on health-related issues and medical help for people with chronic health challenges such as hypertension. The technology enables real-time management by tracking hundreds of metabolic events that occur in the body daily while also reminding the patient to live a healthy lifestyle. The initiative comprises a smart medical solutions system for controlling chronic disease, which makes use of seamless connectivity via the free Wi-Fi access with the city, mobile networks, or laptops. It has also implemented algorithms to aid in the prevention of critical care situations (Chen et al. 2012).

Remote controls are also useful for newborns. Clothing containing respiration sensors "monitors the baby's body posture, activity level, and skin temperature," according to the manufacturer. All of this information is available to parents via an Apple or Google Play store app or, in the future, a light-up smart mug that displays the respiratory patterns of infants (Wollman 2022). Wearable devices, similar to baby monitors, let patients keep track of infant health, and smart diapers check moisture levels and alert parents when diapers should be replaced or when sores are developing.

Doctors can forecast which patients are more vulnerable to particular diseases using predictive modelling. Detailed medical informatics and lifestyle variables can be used to identify people with health or genetic makeup issues. Penn Signals, a study at the University of Pennsylvania Medical School, combines historical and present data to determine which people are at risk for heart failure or sepsis. When patients are discharged from the hospital, nurses receive text messages on their post-discharge care. Patients are to be enlisted into the monitoring programs or specialized care geared to address specific symptoms, depending on their risk profile (Davis 2015).

14.5.3 Data Analytics and Treatment

Trusted data analytics provide tangible benefits in the field of digital medicine. As digital infrastructure improves, the ability to mine health data will expand, allowing doctors and patients to obtain the knowledge they need to take objective decisions. Indeed, being able to carry out real-time evaluation of data will allow for rapid learning about treatment impacts. Physicians can combine and examine information in new and inventive ways using data analysis. They can utilize this data to unearth "actionable insights," learn in real-time, and use what they've learned to select the most successful therapies. When vital signs go outside of permissible ranges, alerts can notify doctors or even patients.

An analytics platform that brings together patient data from multiple sources known as The Collaborative Cancer Cloud enables participants to be able to securely share genetic data of patients, imaging data, and clinical examination data for potentially life-saving discoveries. It will allow vast amounts of data from sites all over the world to be distributed while maintaining the patient's data privacy and security at every location (Vanian 2016).

This has revolutionized cancer treatment and allowed for the fine-tuning of cancer medicines. Researchers can use the cloud platform to make discrete queries regarding specific cancers and receive aggregated data on those patients. The federated approach of the Collaborative allows companies to share de-identified data while maintaining control over their medical data.

Machine learning is crucial to the dynamics of the technologically changing world. Doctors can use artificial intelligence to help them decipher large databases. Wearable gadgets can send out medical notifications if the user experiences an intense situation or medical emergency. It will play a significant role in the development of wireless healthcare systems in the future. There are numerous instances of wearable medical devices or implantable medical sensors, such as cochlear implants, cardiac defibrillators/pacemakers, and insulin pumps, that have a significant influence on patients and provide them with a great deal of comfort. Interfacing these wireless sensing devices with 5G in the future will bring both unparalleled opportunities and formidable hurdles. (Latif et al. 2017). A patient who has fallen and is unable to get up can use such a device to call for assistance. Wearable devices and smartphones have the potential of safeguarding the life of people from going down with a heart attack or stroke by sending an early signal. This device alerts users of the update on medical prescriptions or reminds them of the time to take the nest dosage of medication. Once delay levels are reduced to tiny intervals, remote surgery will be viable. For certain surgeries, surgeons will be able to use virtual tools. Other improvements are already in use; for example, operating robots that aid in small procedures are currently available

The potential of data analytics techniques in healthcare delivery cannot be overemphasized. Health data are complicated and variable, and the current industry standard is to condense it to a diagnostic code. For the patient, the care provider, and the health system, this is a huge loss of information. More can be done with big data analytics offered by 5G. Researchers will have a better understanding of the pharmaceuticals that patients take, how they react to them, and how it all pertains to a specific patient. From the EDGE to the data center, machine learning is integrated, with analytics connecting the two. With the growth in the connected world, a 5G network will enable us to move away from algorithms that are dependent on static data and toward ones that can be tuned in real-time utilizing data from the user (West 2009).

14.6 CONCLUSION

The role of 5G in healthcare centres on the provision of healthcare remotely, through a network of connected devices which involves the exchange of data between multiple connected devices in real-time to bring about efficiency, reduced treatment cost and quick diagnosis. The introduction of 5G brought about a paradigm shift for health services that will enable physicians to pursue precision medicine which is fundamentally different from today's one-size-fits-all model. This chapter presented the evolution of wireless networks from 1G to 6G in detail, along with their strengths and weaknesses. 5G applications in healthcare delivery are discussed, as well as some of the hurdles that may impede their use. The chapter concludes that healthcare centres equipped with 5G wireless technology can considerably increase remote access to medical experts and medical services in real-time. 5G will assist healthcare through enhancing chronic disease management and enabling new highly personalized diagnostic and treatment approaches. It will also enable new, high-value health services for consumers and industrial partners such as Industry 4.0 assisted maintenance services.

REFERENCES

Adetunji, C.O., Michael, O.S., Nwankwo, W., Anani, O.A., Adetunji, J.B., Olayinka, A.S., and Akram, M., 2021. Biogenic nanoparticles based drugs derived from medicinal plants: A sustainable panacea for the treatment of malaria. In: *Green Synthesis in Nanomedicine and Human Health*. CRC Press, 103–122.

Ahad, A., Tahir, M., Sheikh, M.A., Ahmed, K.I., Mughees, A., and Numani, A., 2020. Technologies trend towards 5g network for smart health-care using iot: A review. *Sensors (Switzerland)*, 20 (14).

Alami-Kamouri, S., Orhanou, G., and Elhajji, S., 2019. Mobile agent paradigm integration for new service models in smart ambulance. *International Journal of Innovative Technology and Exploring Engineering*, 8 (6).

Anitha, J., Mahajana, N., and Vellaichamy, S.S., 2015. 4G mobile communications - Emerging technologies. *International Journal of Multidisciplinary Approach and Studies*, 2 (6).

Annis, T., Pleasants, S., Hultman, G., Lindemann, E., Thompson, J.A., Billecke, S., Badlani, S., and Melton, G.B., 2020. Rapid implementation of a COVID-19 remote patient monitoring program. *Journal of the American Medical Informatics Association*, 27 (8).

Ashleibta, A.M., Taha, A., Khan, M.A., Taylor, W., Tahir, A., Zoha, A., Abbasi, Q.H., and Imran, M.A., 2021. 5G-enabled contactless multi-user presence and activity detection for independent assisted living. *Scientific Reports*, 11 (1).

Belikaidis, I.P., Georgakopoulos, A., Demestichas, P., Miscopein, B., Filo, M., Vahid, S., Okyere, B., and Fitch, M., 2017. Multi-RAT dynamic spectrum access for 5G heterogeneous networks: The SPEED-5G approach. *IEEE Wireless Communications*, 24 (5), 14–22.

Brown, G., 2018. Ultra-reliable low-latency 5G for industrial automation. *5G Americas Heavy reading White paper.*

Calton, B., Abedini, N., and Fratkin, M., 2020. Telemedicine in the time of coronavirus. *Journal of Pain and Symptom Management*, 60 (1).

Chen, M., Yang, J., Zhou, J., Hao, Y., Zhang, J., and Youn, C.H., 2018. 5G-smart diabetes: Toward personalized diabetes diagnosis with healthcare big data clouds. *IEEE Communications Magazine*, 56 (4).

Chen, M.J., Chiang, S.J., Lee, J.S., and Yu, E.W.R., 2012. The ctizen telehealth care service model in Taipei: A case study. *2012 IEEE 14th International Conference on e-Health Networking, Applications and Services, Healthcom 2012*, 399–402.

Cheng, L.K., Huang, H.L., and Yang, S.Y., 2021. Attitude toward 5G: The moderating effect of regulatory focus. *Technology in Society*, 67.

Davis, J., 2015. Penn medicine taps big data to save lives [online]. Available from: https://www.informationweek.com/analytics/penn-medicine-taps-big-data-to-save-lives [Accessed 2 Feb 2022].

Desselle, M.R., Brown, R.A., James, A.R., Midwinter, M.J., Powell, S.K., and Woodruff, M.A., 2020. Augmented and virtual reality in surgery. *Computing in Science and Engineering*, 22 (3).

Doraiswamy, S., Abraham, A., Mamtani, R., and Cheema, S., 2020. Use of telehealth during the COVID-19 pandemic: Scoping review. *Journal of Medical Internet Research.* 10.2196/24087

Dutta, G., 2020. Electrochemical biosensors for rapid detection of malaria. *Materials Science for Energy Technologies*. https://doi.org/10.1016/j.mset.2019.10.003

Elsaadany, A., Sedky, A., and Elkholy, N., 2018. A triggering mechanism for end-to-end IoT eHealth system with connected ambulance vehicles. In: *2017 8th International Conference on Information, Intelligence, Systems and Applications, IISA 2017.*

Galazzo, R., 2022. Timeline from 1G to 5G: A brief history on cell phones - CENGN [online]. Available from: https://www.cengn.ca/information-centre/innovation/timeline-from-1g-to-5g-a-brief-history-on-cell-phones/ [Accessed 8 Jan 2022].

Ghazal, T.M., 2021. Positioning of UAV base stations using 5G and beyond networks for IoMT applications. *Arabian Journal for Science and Engineering. 48*, 5687.

Gil, E., Escolà, A., Rosell, J.R., Planas, S., and Val, L., 2007. Variable rate application of plant protection products in vineyard using ultrasonic sensors. *Crop Protection.* https://doi.org/10.1016/j.cropro.2006.11.003

Gilbert, A.W., Mentzakis, E., May, C.R., Stokes, M., and Jones, J., 2021. Patient preferences for use of virtual consultations in an orthopaedic rehabilitation setting: Results from a discrete choice experiment. *Journal of Health Services Research and Policy.* 10.1177/13558196211035427

Gogia, S., 2020. Rationale, history, and basics of telehealth. *Fundamentals of Telemedicine and Telehealth*, 11–34. https://doi.org/10.1016/B978-0-12-814309-4.00002-1

Gordon, W.J., Henderson, D., Desharone, A., Fisher, H.N., Judge, J., Levine, D.M., MacLean, L., Sousa, D., Su, M.Y., and Boxer, R., 2020. Remote patient monitoring program for hospital discharged COVID-19 Patients. *Applied Clinical Informatics*, 11 (5).

Griggs, K.N., Ossipova, O., Kohlios, C.P., Baccarini, A.N., Howson, E.A., and Hayajneh, T., 2018. Healthcare blockchain system using smart contracts for secure automated remote patient monitoring. *Journal of Medical Systems*, 42 (7).

Hameed, K., Bajwa, I.S., Sarwar, N., Anwar, W., Mushtaq, Z., and Rashid, T., 2021. Integration of 5G and block-chain technologies in smart telemedicine using IoT. *Journal of Healthcare Engineering*, 2021.

Haverans, R., 2022. A quick history on 1G to 5G – A 40-year journey. *Apis* [online]. Available from: https://apistraining.com/from-1g-to-5g/ [Accessed 8 Jan 2022].

Hu, J., Liang, W., Hosam, O., Hsieh, M.Y., and Su, X., 2021. 5GSS: A framework for 5G-secure-smart healthcare monitoring. *Connection Science*. https://doi.org/10.1080/09540091.2021.1977243

Huang, J., Chang, Q., Zou, J., and Arinez, J., 2018. A real-time maintenance policy for multi-stage manufacturing systems considering imperfect maintenance effects. *IEEE Access*, 6.

Hume, R. and Jeff, L., 2022. Designing for telemedicine spaces | HFM | Health Facilities Management [online]. Available from: https://www.hfmmagazine.com/articles/1889-designing-for-telemedicine-spaces [Accessed 2 Feb 2022].

Kamruzzaman, M.M., 2020. Architecture of smart health care system using artificial intelligence. In: *2020 IEEE International Conference on Multimedia and Expo Workshops, ICMEW 2020*.

Kiesel, R., van Roessel, J., and Schmitt, R.H., 2020. Quantification of economic potential of 5G for latency critical applications in production. In: *Procedia Manufacturing*.

Kikuchi, T., Tsujimura, K., and Kato, K., 1985. Progress in the 800 MHz land mobile TELEPHONE system in Japan. In: *IEEE Vehicular Technology Conference*.

Kilvert, A., Wilmot, E.G., Davies, M., and Fox, C., 2020. Virtual consultations: are we missing anything? *Practical Diabetes*, 37(4).

Kumari, A., Kumar, V., Abbasi, M.Y., Kumari, S., Chaudhary, P., and Chen, C.M., 2020. CSEF: Cloud-based secure and efficient framework for smart medical system using ECC. *IEEE Access*, 8.

Latif, S., Qadir, J., Farooq, S., and Imran, M.A., 2017. How 5G wireless (and concomitant technologies) will revolutionize healthcare? *Future Internet*, 9(4), 1–24.

Laukka, E., Pölkki, T., Heponiemi, T., Kaihlanen, A.M., and Kanste, O., 2021. Leadership in digital health services: Protocol for a concept analysis. *JMIR Research Protocols*, 10 (2).

Lee, A.K., Jeon, S.B., and Choi, H. Do, 2021. EMF levels in 5G new radio environment in Seoul, Korea. *IEEE Access*, 9, 19716–19722. 10.1109/ACCESS.2021.3054363

Li, G., Lian, W., Qu, H., Li, Z., Zhou, Q., and Tian, J., 2021. Improving patient care through the development of a 5G-powered smart hospital. *Nature Medicine*, 27, 936–937

Li, P., Luo, Y., Yu, X., Wen, J., Mason, E., Li, W., and Jalali, M.S., 2020. Patients' perceptions of barriers and facilitators to the adoption of e-hospitals: Cross-sectional study in western China. *Journal of Medical Internet Research*, 22 (6).

Lian, W., Wen, L., Zhou, Q., Zhu, W., Duan, W., Xiao, X., Mhungu, F., Huang, W., Li, C., Cheng, W., and Tian, J., 2020. Digital health technologies respond to the COVID-19 pandemic in a tertiary hospital in China: Development and usability study. *Journal of Medical Internet Research*, 22 (11).

Liu, X., Li, N., Liu, Y., and He, Y., 2021. *Smart medical and nursing platform based on 5G technology*, 285–295.

Malasinghe, L.P., Ramzan, N., and Dahal, K., 2019. Remote patient monitoring: a comprehensive study. *Journal of Ambient Intelligence and Humanized Computing*, 10 (1).

Melton, S., Simmons, S.C., Smith, B.A., and Hamilton, D.R., 2020. Telemedicine. In: *Principles of Clinical Medicine for Space Flight*.

Mishra, L., Vikash, and Varma, S., 2021. Seamless health monitoring using 5G NR for Internet of Medical Things. *Wireless Personal Communications*, 120 (3).

Mourtzis, D., Siatras, V., and Angelopoulos, J., 2020. Real-time remote maintenance support based on augmented reality (AR). *Applied Sciences (Switzerland)*, 10 (5).

Muñoz-Saavedra, L., Miró-Amarante, L., and Domínguez-Morales, M., 2020. Augmented and virtual reality evolution and future tendency. *Applied Sciences (Switzerland)*, 10 (1).

Murthy, V., Herbert, C., Bains, D., Escudier, M., Carey, B., and Ormond, M., 2021. Patient experience of virtual consultations in oral medicine during the COVID-19 pandemic. *Oral Diseases*. 10.1111/odi.14006

Ndinojuo, B.-C.E., 2020. 5G, Religion, and misconceptions in communication during Covid-19 in Nigeria. *Jurnal The Messenger*, 12 (2).

Newswire, P.R., 2016. Global Internet of Things (IoT) industry. *Lon-Reportbuyer*.

Nguyen, P.T., Lorate Shiny, M., Shankar, K., Hashim, W., and Maseleno, A., 2019. Robotic surgery. *International Journal of Engineering and Advanced Technology*, 8 (6 sSpecial Issue 2).

Noura, M., Atiquzzaman, M., and Gaedke, M., 2019. Interoperability in Internet of Things: Taxonomies and open challenges. *Mobile Networks and Applications*, 24 (3).

Nwankwo, W., Olayinka, A.S., and Ukhurebor, K.E., 2020. Nanoinformatics: Why design of projects on nanomedicine development and clinical applications may fail? In: *2020 International Conference in Mathematics, Computer Engineering and Computer Science, ICMCECS 2020*.

O'Regan, G., 2016. A short history of telecommunications.

Olayinka, T. C., Olayinka, A.S., and Oladimeji, O.R., 2018. Framework for web-based and desktop-based telemedicine system. *Journal of Computer Engineering*. 20, 1

Olayinka, T.C., Aladeselu, V.A., and Olayinka, A.S., n.d. Architectural framework for web-based telemedicine system.

Olayinka, T.C. and Chiemeke, S.C., 2019. Predicting paediatric malaria occurrence using classification algorithm in data mining. *Journal of Advances in Mathematics and Computer Science*, 1–10. 10.9734/jamcs/2019/v31i430118

Padmashree, T. and Nayak, S.S., 2020. 5G technology for e-health. In: *Proceedings of the 4th International Conference on IoT in Social, Mobile, Analytics and Cloud, ISMAC 2020*.

Paterson, C., Bacon, R., Dwyer, R., Morrison, K.S., Toohey, K., O'Dea, A., Slade, J., Mortazavi, R., Roberts, C., Pranavan, G., Cooney, C., Nahon, I., and Hayes, S.C., 2020. The role of telehealth during the COVID-19 pandemic across the interdisciplinary cancer team: Implications for practice. *Seminars in Oncology Nursing*. 10.1016/j.soncn.2020.151090

Prevention, C.-C. for D.C. and, 2021. CDC - Malaria - malaria worldwide - impact of malaria. 10.1016/j.soncn.2020.151090

Proulx-Cabana, S., Segal, T.Y., Gregorowski, A., Hargreaves, D., and Flannery, H., 2021. Virtual consultations: Young people and their parents' experience. *Adolescent Health, Medicine and Therapeutics*, 12.

Rath, M. and Pattanayak, B., 2019. Technological improvement in modern health care applications using Internet of Things (IoT) and proposal of novel health care approach. *International Journal of Human Rights in Healthcare*, 12 (2).

Sabah, N., 2016. 4G technology and its applications [2] magic of 4G technology. *International Journal of Research in Engineering, Technology and Science*, VI (July).

Sachs, J., Andersson, L.A.A., Araujo, J., Curescu, C., Lundsjo, J., Rune, G., Steinbach, E., and Wikstrom, G., 2019. Adaptive 5G low-latency communication for tactile internet services. *Proceedings of the IEEE*. 107, 2

Saravia, A., 2022. The evolution of wireless telecommunication: From 1G to 5G [online]. Available from: https://www.ufinet.com/the-evolution-of-wireless-telecommunication-from-1g-to-5g/ [Accessed 8 Jan 2022].

Selmaoui, B., Mazet, P., Petit, P.B., Kim, K., Choi, D., and de Seze, R., 2021. Exposure of South Korean population to 5G mobile phone networks (3.4–3.8 GHz). *Bioelectromagnetics*, 42 (5), 407–414.

Setyo, J., Gunawan, A., Sani, M.R., and Wang, G., 2019. Challenge of 5g network technology for telemedicine and telesurgery. *International Journal of Advanced Trends in Computer Science and Engineering*, 8 (6), 3680–3683.

Shu, M. and Li, J., 2020. Health digital technology in COVID-19 pandemic: Experience from China. *BMJ Innovations*.

Singh, R.P., Javaid, M., Kataria, R., Tyagi, M., Haleem, A., and Suman, R., 2020. Significant applications of virtual reality for COVID-19 pandemic. *Diabetes and Metabolic Syndrome: Clinical Research and Reviews*, 14 (4).

Sinha, V., Malik, M., Nugent, N., Drake, P., and Cavale, N., 2021. The role of virtual consultations in plastic surgery during COVID-19 lockdown. *Aesthetic Plastic Surgery*, 45 (2).

Siriwardhana, Y., Gür, G., Ylianttila, M., and Liyanage, M., 2021. The role of 5G for digital healthcare against COVID-19 pandemic: Opportunities and challenges. *ICT Express*, 7 (2), 244–252.

Tarikere, S., Donner, I., and Woods, D., 2021. Diagnosing a healthcare cybersecurity crisis: The impact of IoMT advancements and 5G. *Business Horizons*, 64 (6).

Tian, S., Yang, W., Grange, J.M. Le, Wang, P., Huang, W., and Ye, Z., 2019. Smart healthcare: making medical care more intelligent. *Global Health Journal*, 3 (3), 62–65.

Ullah, S., Kim, K. Il, Kim, K.H., Imran, M., Khan, P., Tovar, E., and Ali, F., 2019. UAV-enabled healthcare architecture: Issues and challenges. *Future Generation Computer Systems*, 97.

Usman, M.A., Philip, N.Y., and Politis, C., 2019. 5G enabled mobile healthcare for ambulances. In: *2019 IEEE Globecom Workshops, GC Wkshps 2019 - Proceedings*.

Vanian, J., 2016. Intel's cancer cloud gets new recruits [online]. Available from: https://au.finance.yahoo.com/news/intel-cancer-cloud-gets-recruits-215623014.html?guccounter=1&guce_referrer=aHR0cHM6Ly93d3cuZ29vZ2xlLmNvbS8&guce_referrer_sig=AQAAAA3v5NB_mPms4oV2sQ5XxXJ0NHriRmJpLPMJQhspiDR7TZcuOvSI1mCS3oMVNkjK6CUTeruR0eqLAb-c7L0DPMiw-B8TCp3-jTF86TQTgqfnWmkk0XJP4rJSwLAjfvUpSyS6DhN19K8I66WMyQDTJz-sZIox7bDPvXxdk4-OSUbt [Accessed 2 Feb 2022].

West, D.M., 2009. *How 5G technology enables the health Internet of Things*. Cyber Resilience of Systems and Networks.

WHO, 2020. *WHO World Malaria Report 2020*. Malaria report.

Wollman, D., 2022. The internet of toddlers: Intel shows off a smart baby onesie. *Engadget* [online]. Available from: https://www.engadget.com/2014-01-07-intel-smart-baby-onesie.html?guccounter=1 [Accessed 2 Feb 2022].

Wong, C.K., Ho, D.T.Y., Tam, A.R., Zhou, M., Lau, Y.M., Tang, M.O.Y., Tong, R.C.F., Rajput, K.S., Chen, G., Chan, S.C., Siu, C.W., and Hung, I.F.N., 2020. Artificial intelligence mobile health platform for early detection of COVID-19 in quarantine subjects using a wearable biosensor: Protocol for a randomised controlled trial. *BMJ Open*. https://doi.org/10.1136/bmjopen-2020-038555

Ying, S. and Yu, W., 2021. Application of 5G technology in the construction of intelligent health management system. In: *Communications in Computer and Information Science*.

Index

Pages in *italics* refer to figures and pages in **bold** refer to tables.

R

S

T

V

W

Printed in the United States
by Baker & Taylor Publisher Services